AF522280

Genetic Diversity and Phenotypic Stability in Crop Plants

nipa
Browse Subject
Agriculture Sciences (679)
Animal Husbandry and Veterinary Sciences (261)
Climate Change and Environmental Sciences (63)
Community Science (Home Science) (39)
Competitive Examinations and MCQ's (57)
Dairy Technology (91)
Fisheries Sciences (26)
Food Science and Technology (155)
Geological Sciences (50)
Horticultural Sciences (342)
Plant Sciences (191)
Skill and Entrepreneurship Development (47)
eBooks / eChapters / Articles
eBooks
Explore Now
An e book is an electronic book available in various digital formats, can be accessed from anywhere, anytime on desktops/laptops/smart phones and tablets connected to internet.
Explore Now
eBooks / eChapters / Articles
Browse, Search, Read & Buy...
Subject Catalogues
eChapters
Publishers
Print Books
Forthcoming
eBooks
New Titles
scan for catalogue
NIPA
GENX
ONLINE RESOURCES
Publishing Books and Journals
Ebooks and e articles, Current Affairs
Reasoning, Logic and Aptitude
Competitive Examination Preparation
Language Learning and Development Programme
Document Quality Checker and Improvement Tool
Effective Public Speaking, Presentation and Interpersonal Skills
Online Programmes for Professional Development
Personality Development and Human Values
PAY USING
UPI
PayPal

Genetic Diversity and Phenotypic Stability in Crop Plants

K. Manivannan
S. Thirugnanakumar
P. Thangavelu
K. Saravanan
N. Senthilkumar
A. Anandan

NEW INDIA PUBLISHING AGENCY
Pitam Pura, New Delhi – 110 088

NEW INDIA PUBLISHING AGENCY
101, Vikas Surya Plaza, CU Block, LSC Market
Pitam Pura, New Delhi 110 034, India
Phone: + 91 (11)27 34 17 17 Fax: + 91(11) 27 34 16 16
Email: info@nipabooks.com
Web: www.nipabooks.com

Feedback at feedbacks@nipabooks.com

ISBN : 978-93-85516-95-5

Composed, Designed and Printed in India

Preface

Genetic diversity plays an important role in crop evolution. Assessing diversity by morphometric through classical plant breeding is very much useful to study the crop evolution and to identify suitable parents for developing superior progenies. The genetically diverse parents will show high combining ability resulting in high heterosis in F_1 generation and throw some useful segregants in the F_2 and in later segregating generations too. Genetic diversity has been classically measured with Mahalanobi's' D^2 method. Diversity analysis over season and environments also gives useful information in identification of suitable parents. The consistent genetic diversity over seasons and environments would be useful in practical plant breeding. Further, genetic diversity has also being measured with molecular markers to avoid environmental interaction.

On the other hand, phenotypic stability among genotypes is very important in actual plant breeding. Among the various methods adopted to measure phenotypic stability, the method outlined by Eberhart and Russell gains momentum, as it is simple and explicit in nature. Phenotypic stability parameters has been collectively utilized to classify the genotypes as having high stable genotypes with average responsiveness as well as genotypes suitable for favourable as well as unfavourable environments.

The authors have taken concerted efforts to bring useful information on genetic diversity and phenotypic stability among various field and horticultural crops for the researches who involved in crop improvement program. Further, the authors welcome suggestions from readers to improve this book further.

Authors

Contents

1

Mahalanobi's D^2 Statistic

Genetic architecture of a population is the result of prolonged natural selection. The populations which exist in diverse environments might have been strongly diversified genetically. So, it is necessary to understand the extent and rates of genetic divergence existing between the diversified forms or types. Genetic diversity must be sufficient enough to justify breeding effort. Genetic diversity may be available in improved germplasm or it may exist only in genetically inferior stocks. The availability of statistical tools to quantitatively measure the genetic divergence between two or more populations and the relative contribution of individual characters to the total divergence have permitted to trace the evolutionary patterns in some crops such as rice and tobacco and in choosing the parents for hybridization in crop plants (Rao, 1958; Blackith, 1960; Morishima and Oka, 1960; Murthy *et al.,* 1962). Among the several statistical methods developed for measuring the divergence between population, multivariate analysis (D^2 statistic) developed by Mahalanobis in 1936, has been found to be potent tool.

Mahalanobis D^2 analysis is an effective tool in quantifying the degree of divergence at genetic level and provides a quantitative measures of the association between geographic and genetic diversity based on generalized distance.

Geographical distribution was considered as an index of genetic diversity by many workers in the past. Jawahar Ram and Panwar (1970) in rice and Griffing and Lindstorm (1954) and Dhawan and Singh (1961) in maize observed the parallelism between genetic diversity and geographic distribution. However, absence of such parallelism and presence of forces other than geographical distribution being responsible for genetic divergene were brought out by Trehan *et al.* (1974), Govinda Raju (1982) and Thirugnanakumar (1991) in sesame.

Genetic divergence

Before taking up the study of genetic divergence, test of significance of difference with regard to the pooled effect of all the characters using Wilk's lambda criterion has to be carried out.

Wilk's criterion

This will be estimated using the formula given by Wilks (Rao, 1952)

Wilks ^ criterion = $\frac{[E]}{[E+V]}$

Where [E] = Determinant of error sum of squares and sum of product matrix

[E + V] = Determinant of (genotype + error) sum of squares and sum of product matrix

Then the value of 'V' statistic will be worked out as follows (Singh and Choudhary, 1985).

'V' Stat = - m $\log_e$

Where, m = $n - \frac{[p+q+q]}{[2]}$

P = Number of characters

Q = Number of genotypes –1

N = Degrees of freedom for error + degrees of freedom for genotypes

The value of 'V' statistic was compared with the tabulated chi-square value for pq degrees of freedom and thus significance was tested.

D^2 analysis

As suggested by Mahalanobis (1936), D^2 statistic is used for estimating the genetic divergence among the genotypes studied. The data is subjected to multivariate analysis (Rao, 1952; Method-B1). Group constellations of clusters is being done as outlined in Singh and Choudhary (1985).

Intra and inter cluster distance

After establishing the clusters, the intra-cluster distance is worked out by taking the average of the component genotypes in that cluster. The average inter-cluster divergence is arrived at by taking into consideration of all the component D^2 values possible among the members of the two clusters considered. The square root of the average D^2 values give the genetic distance 'D' between the clusters.

Based on D values (inter-cluster) the following scale for rating of the distance is adopted (Rao, 1952).

Category	D Values
Less divergent (L)	99 and below
Moderately divergent (M)	Between 100 and 200
Highly divergent (H)	Above 200

Based on D^2 values (inter-cluster distance), the following scale for rating of the distance is also adopted (Arunachalam and Bandyopadhyay, 1984) by calculating the mean (m) and standard deviation (S) of the values of divergence. Four divergence classes are defined as follows:

Category	Divergence class
DC1	$D^2 \geq (m + s)$
DC2	$D^2 < (m + s)$ and $\geq m$
DC3	$D^2 \geq (m - s)$ and $< m$
DC4	$D^2 < (m - s)$

2

Genetic Diversity in Rice [*Oryza sativa* (L.)]

Thirugnanakumar, S. and R. Balamurugan

Rice (2n = 24) is an annual grass (family: Graminae) with terminal panicles bearing the grains. It originated in Asia between the fourth and fifth millennia B.C. A substantial portion of the people in south and Southeast Asia and the southern part of China are depending on rice as staple food (Chang, 1976). Rice is planted in the most diverse environmental conditions of any major food crop, from sea level to around 3000 m altitude and from 30°S to 49°N. Temperature and day lengths are therefore quite diverse in the rice growing areas. Rice is also grown under water regimes ranging from dry land to deepwater fields where water may reach 5 m depth. The wide geographical distribution of rice and its early history of cultivation have resulted in the development of a great diversity of varietal types (Vergara, 1985).

Rice has been used as food by man for over 10,000 years and has fed a great number of people for a longer period than has any other crop. "Rice is life" for almost four billion people of the world contributing over 20 per cent of the total caloric intake of human population (Chaudhary and Tran, 2001). Rice is cultivated in 113 countries and on all continents except Antarctica and it is the staple food for 17 countries in Asia and the pacific, eight countries in Africa, seven countries, in Latin America and the Caribbean and one in the near east (IRC, 2003).

Globally rice is grown in around 155 million ha occupying one tenth of arable land, with an annual production of 579 million tonnes of rough rice (Iiyas Ahmed, 2003). The average productivity of 3.89 t ha^{-1} of rice provides a per capita. Consumption of 100 to 240 kg per year across the globe. Over 90 per cent of the world's rice is produced and consumed in Asia and more than half of it from India and China exploiting maximum arable land. More than two billion people obtain their 80 per cent calorie from rice and its derived products from 90 per cent of world's rice area (FAO, 1999). Rice based on production systems and their associated post harvest operations employ nearly a billion people in rural areas in developing countries (IRC, 2003).

In India, rice area accounts 22.8 per cent of total cropped area, 35.6 per cent of area under food crop and 43.9 per cent of area under cereals, while rice production accounts 31 per cent of food grain production and 45 per cent of cereal production providing 20 to 25 per cent agricultural income. The rice production has increased from 20 million tonnes in 1950-1951 to 91.6 million tonnes in 2001-2002 with a productivity increase of 0.6 to 2.2 t ha^{-1}, during the same period. There has been a substantial increase in area from 30 million ha to 45.5 million ha from 1950 to 2001 (Singh, 2002). This has been due to crop substitution and increased profitability by growing semi-dwarf high yielding varieties under irrigated conditions.

At present the population in India is 1130 million, which is increasing by 17 million a year (Singh, 2002). To meet the food demand of growing population, five million tonnes of additional food is required, out of which, two million share is of rice. Hence, rice demand of 143 million tonnes by 2030 has to be met by increased rice productivity per unit time and area, as there is little scope for increase in area in the future. To meet the increasing demand, rice production must be increased inspite of less land, less water and less pesticides in a sustainable way. Stability of yield over environments, season as well as locations remains to be one of the important objective of rice breeders. Particularly, stability rice under physical and physiological stress is important, as water is a precise and valuable input. It becomes the earnest duty of rice breeder to find out the rice genotypes suitable for aerobic and saline conditions to fit than in to sustainable agriculture.

Water is a crisis in many parts of the world. India is also experiencing shortage of water for agriculture. When water is used for agricultural purposes, scientific water management seldom goes into practice. There has been 35 fold increase in the water use during the last three centuries. During this century, global water withdrawal rate has been twice the rate of population growth rate. This has resulted in 40-50 per cent reduction in per capita availability of water in Asia during 1955-1990 period. Agriculture still remains the biggest consumer of water using 3300 cubic kilometers of water per annum. This is six times more than the industrial use. In Asia, nearly 86 per cent of total annual withdrawal of water goes for agriculture. One of the factors associated with variation in water requirement is the system of cultivation and duration of crop. As an example, computation of water requirement for Regional Agricultural Research Station, Pattambi, Kerala, shows that dry broadcast crop 130 days duration needed 1116 mm of water, while a transplanted crop of 100 days duration, needed, 1213 mm of water (Kurien, 1982). Dry seeding of crops on an average consumers only 700-900 mm water, which means that we can plant an early duration crop based on the reservoir capacity (IRRI, 1995).

Water saves life and so becomes precious. It will become scarce in due curse of time. So, rice breeders like other breeders, has to find out a way to grow a successful rice crop with limited water source.

De and Rao (1987) observed the data on grain yield and eight related traits from thirty rice varieties grown in deep water (750 cm) for entire growing season. Using the Mahalanobis D^2 statistic, the varieties could be grouped into ten clusters. They also reported that the geographical distribution and genetic divergence were not associated.

Pande and Ghora (1987) studied the genetic divergence of forty eight improved varieties of cultivated rice using D^2 analysis and observed that grouping of genotypes in the same cluster confirmed that there was no parallelism between genetic diversity and geographical distribution.

Fifty lowland rice cultivars were grouped into ten clusters using D^2 analysis. Cluster, I, II and III were the largest and together accounted for more than two thirds of total population. Genotypic diversity was not related to geographical diversity (Singh *et al.*, 1987).

Genetic divergence study on twenty five breeding lines of rice from diverse pedigree, undertaken in two levels of water regimes (shallow and intermediate water) revealed six and seven clusters of varying constellation (Pradhan and Roy, 1990). Anandakumar and Subramanian (1992) studied genetic divergence in twenty three cultivars of rice. It was found that geographical diversity need not necessarily be related to genetic divergence. Locally adapted diverse genotypes like E 45 Moongil Samba and Arurakkari might serve as better parents for upland rice improvement.

A quantitative assessment of the genetic divergence for nine genotypes of rice over three environments by multivariate analysis using Mahalanobis D^2 analysis was carried out by Ibrahim *et al.* (1992). They grouped the nine genotypes into three clusters in E_1, E_2 and E_3. Difference in environment did not adversely affect the clustering pattern bearing minor adjustments between groups.

Roy and Panwar (1993) studied genetic divergence on ninety-nine diverse genotypes of rice using multivariate D^2 analysis. Genotypes were grouped into sixteen clusters. Genetic divergence was controlled mainly by number of panicles per plant, number of grains per panicle, grain yield per plant and bacterial blight severity.

The nature and magnitude of genetic divergence was assessed in twenty eight genotypes of rainfed rice using Mahalanobis D^2 statistic (Vivekanandan and Sukanyasubramaniyan, 1993). The population was grouped into five clusters.

The geographical diversity was not related to genetic diversity.

Genetic diversity among forty four breeding lines and two improved cultivars were studied under rainfed upland condition using Mahalanobis D^2 clusters. The cluster with twelve genotypes was the largest, while clusters XI and XII were monogenotypic. Genotypes in cluster IX had the highest intra – cluster distance. The maximum inter cluster distance (3920) was observed between genotypes of clusters VII and XII (Chauhan and Chauhan, 1994).

Mishra *et al.* (1994) studied multivariate analysis of divergence for seven quantitative traits among thirty seven strains of rice. They grouped the genotypes into five clusters. The first cluster contained thirty one genotypes, the second contained three, while the third, fourth and fifth clusters contained one genotype each.

Sarawgi and Shrivastava (1996) grouped sixteen parents and their seventy two cross combinations based on D^2 values of fifteen characters under irrigated and rainfed conditions. The genotypes were grouped into seven and fifteen clusters respectively under irrigated and rainfed conditions.

Seventy five genotypes of rice were evaluated by Sawant *et al.* (1996) for genetic divergence on eight yield components and were grouped into ten clusters. The average inter-cluster distance was the highest between clusters IX and X (66.58) followed by clusters VI and IX (62.59).

Sarathe *et al.* (1996) studied genetic diversity and heterosis in sixty five rice varieties. They were grouped into eighteen clusters. They suggested parents having fairly high to medium diversity estimates might be utilized in hybridization programme in order to increase the chances of getting high heterotic manifestation.

Rao and Gomathinayagam (1997) studied the genetic divergence in forty rice genotypes planted under semi – dry condition at two different locations. Differential response of the genotypes to the environments had altered the clustering pattern between environments. Favourable environment yielded in lesser number of clusters by exploiting high genetic potential of different traits expressed by genotypes.

Sarma *et al.* (1997) collected thirty nine rice genotypes of extra-early duration from Madhya Pradesh. The thirty-nine genotypes were grouped into ten and eight clusters under rainfed upland and irrigated lowland conditions, respectively. Although, these genotypes were not eco-geographically diverse, they showed considerable genetic diversity on D^2 analysis. The maximum inter-cluster distance under rainfed upland condition was recorded between clusters V and VI (109.11). On the other hand, under irrigated low land situation, distance between clusters III and VI (267.56) was maximum.

Hanamaratti *et al.* (1998) evaluated fifty rice genotypes for ten yields component in low and upland environments. Cluster analysis grouped the genotypes into eighteen and seventeen clusters under low and upland conditions, respectively. Genetic diversity was not related to geographic diversity.

Genetic divergence was assessed in thirty four rice stocks using D^2 statistic by Bansal *et al.* (1999). The genotypes were grouped into fifteen clusters. Distribution of genotypes within various clusters was independent of geographical distribution.

Kandhola and Panwar (1999) studied genetic diversity among fifty-two indigenous and exotic genotypes of rice using Mahalanobis D^2 statistic. The maximum inter-cluster distance was observed between genotypes of clusters V and XI (18984.00). It is concluded that hybridization among genotypes drawn from widely divergent clusters would be useful.

Soni *et al.* (1999) observed a highly significant variability among one hundred and thirty two genotypes for the quality traits studied. Based on the relative magnitude of D^2 values, that genotypes were grouped into ten clusters. Cluster I had 78 genotypes, while clusters II and III had 17 and 22 genotypes respectively. Clusters IV and V had four genotypes each. Clusters VI and VII has two genotypes each and the remaining three clusters VIII, IX and X had single genotype each. Inter-cluster distance was maximum between cluster VII and VIII (33.19), followed by clusters VI and VIII (32.99).

Hedge and Patil (2000) measured the genetic divergence using Mahalanobis D^2 statistic in forty genotypes of rainfed rice and they grouped the genotypes into seven clusters. The average inter-cluster D^2 value was the highest (51.88) between the clusters V and VII indicating high genetic divergence between the cultivars of these clusters.

Rather *et al.* (2001) studied the genetic diversity in fifty six rice genotypes for twelve characters. They grouped all genotypes in to six clusters and diversity of genotypes was independent of their geographical distribution.

Shanthi and Singh (2001) studied genetic diversity among seventeen induced mutant genotypes of rice using Mohalanobis D^2 statistic and Touches method. The maximum inter cluster distance was observed between genotypes of clusters II and III. It was concluded that hybridization among genotypes drawn from widely divergent clusters would be useful.

Arun *et al.* (2002) grouped one hundred aromatic rice genotypes based on D^2 values. Twenty eight yield and morphological characters were studied. The genotypes grouped into nine clusters.

Chaudhary and Sarawgi (2002) evaluated fifty four rice genotypes for nineteen morphological and quality traits. The genotypes were grouped into five clusters. The maximum inter cluster distance was observed in cluster III. Cluster I had eighteen genotypes, cluster III had sixteen genotypes and cluster V had nine genotypes.

Karthikeyan (2002) studied the genetic divergence in forty two rice genotypes at three different locations. Cluster analysis grouped the genotypes into ten, seven and six clusters under three locations, respectively. D^2 analysis showed the absence of parallelism between geographical origin and genetic diversities.

Reddy *et al.* (2002) studied genetic diversity among thirty six genotypes of rice using Mahalanobis D^2 statistic. The genotypes were grouped into 12 clusters. The maximum inter cluster distance was observed between genotypes of clusters III and VII followed by clusters IX and VI.

Babu *et al.* (2003) measured the genetic divergence using Mahalanobis D^2 statistic in thirty three genotypes and they were grouped the genotypes into ten clusters. The cluster I contained seven genotypes followed by clusters II and III with six genotypes. Clusters VI, IX and X had one genotype each.

Shiv *et al.* (2003) studied genetic diversity among sixty one basmati rice genotypes using Mahalanobis D^2 statistic. The genotypes were grouped into four clusters. The cluster I contained forty three genotypes followed by cluster II with thirteen genotypes, cluster III had four genotypes and cluster IV had only one genotype. The maximum inter cluster distance was observed between genotypes of clusters III and IV.

Shukla and Pandey (2003) evaluated genetic diversity among thirty four thermo sensitive male sterile lines and grouped the genotype into three clusters. Cluster I had 31 genotypes, cluster II and 2 genotypes and cluster III and one genotype. Cluster II showed maximum intra cluster distance and the maximum inter cluster distance was observed between clusters I and II.

Venkatesan (2004) studied genetic diversity among 46 rice genotypes using Mahalanobis D^2 statistics. The genotypes were grouped into six clusters. The cluster I contained 39 genotypes. Cluster II and III with two genotypes each and clusters IV, V and XI with one genotype each. Cluster II showed maximum intra-cluster distance and the maximum inter-cluster distance observed between clusters I and VI.

Sree Krishna (2005) studied genetic diversity among 57 genotypes using Mahalanobis D^2 statistics. The genotypes were grouped into ten clusters. Cluster I comprised 26 genotypes, cluster II having 14 genotypes, cluster II have five genotypes, cluster IV have four genotypes, cluster V have three genotypes and clusters VI, VII, VIII, IX and X with only one genotype each. Cluster V showed maximum intra-cluster distance and maximum inter-cluster distance was found between cluster II and X.

Chaturvedi and Maurya (2005) genetic divergence analysis in 26 genotypes of rice belonging to 7 ecotypic situations. The genotypes were grouped in 8 clusters. Comparison of cluster means revealed that cluster VIII gave exceptionally high values for 7 characters followed by cluster VII. The maximum inter-cluster D^2 value was obtained between cluster III and VI and III and VIII. It is suggested that for developing better cultivar, the genotypes of cluster III could be utilized in hybridization programme with the genotypes of cluster VI and VIII.

Gahalain (2006) studied genetic divergence among 55 rice genotypes grown in Kumaun Himalaya using Mahalanobis D^2 statistic. The 55 genotypes were grouped into twelve clusters. A maximum of 15 genotypes were included in cluster III, followed by cluster II and IV with 9 genotypes each. Dispersion pattern of genotypes over a large number of clusters with a maximum of 15 genotypes in cluster III to five single genotype clusters from VIII to XII indicated the presence of high degree of genetic divergence. Intra-cluster divergence was maximum in cluster VII, which is not significantly different from other clusters, inter-cluster distance was noticed between cluster XII and VII.

Rajasekaran (2006) studied genetic divergence among 50 genotypes based on Mahalanobis D^2 statistic. The genotypes were categorized into six clusters. The intra-cluster distance was the maximum in cluster I and it was the most heterogeneous group. The maximum inter-cluster distance was observed between cluster V and VI. The clusters IV, V and VI recorded maximum cluster mean for grain yield and other yield contributing characters.

Anandan et al. (2011) studied 44 salt tolerant rice genotypes from different geographic regions were assessed using Mahalanobis D^2 and principal component analysis. The D^2 statistics grouped the genotypes into 12 distinct clusters. Consisting of 19 genotypes, Cluster I was the largest cluster, followed by Cluster II with nine genotypes. Meanwhile, Clusters III, IV, VI, VII, IX, and XII were mono-genotypic clusters. The maximum intercluster distance was exhibited between Cluster IX and X (144.91), followed by Clusters II and

X (131.87), as well as clusters VII and X (126.27). Both D^2 and PCA revealed that the morphometric diversity was based on the pedigree and independent of geographical origin. Hybridization among the genotypes which had the maximum inter-cluster distances could produce heterotic combinations and wide variability in segregating generations for many beneficial traits.

Many workers have clarified that the yielding ability of crop plants as a polygenic character, which can be improved by breeding. However, it is a complex character, which is composed of several characters known as field components. Yield per unit area in rice is decided by number of productive tillers per unit area, number of grains per panicle and average grain weight. The highest yield is obtained when the product of these three components are maximum (Grafius, 1956).

It has long been recognized that plants respond to changes in the environmental conditions during the process of growth and development. The yield and its components being quantitative in nature and more sensitive to environmental fluctuation, exhibit differential reaction in their phenotypic expression and also variation in the relationship among them, indicating the problem of direct selection for yield to obtain expected results.

Balamurugan (2007) studied 25 genotypes in rice to assess the genetic diversity of 25 rice genotypes for grain yield and its component characters under three different environments. The results are discussed here under.

Analysis of variance

The analysis of variance for individual as well as pooled analysis revealed significant difference among all the traits of interest. It indicated that there exists sufficient amount of variability among the 25 genotypes of the present study. Therefore, further analysis is appropriate. Similar findings were earlier reported by Rao and Gomathinayagam (1997), Karthikeyan (2002) and Venkatesan (2004).

Mean performance

In E_1, ten genotypes exceeded the population mean grain yield. The genotypes *viz.,* G_{24} and G_{11} recorded the maximum grain yield per plant (34.59 g and 33.69 g, respectively). In E_2 fourteen genotypes exceeded the grand mean grain yield. The genotypes *viz.,* G_{22} and G_1 registered maximum grain yield (32.52 g and 30.24 g, respectively). In E_3, fifteen genotypes out yielded the population mean. The genotypes *viz.,* G_{24} and G_{16} showed maximum grain yield per plant (31.29 g and 27.76 g, respectively). In pooled analysis, 14 genotypes exhibited higher grain yield than the population mean. The genotypes *viz.,* G_{24} and G_{11} registered maximum grain yield per plant (31.10 g and 30.18 g respectively). The genotype namely, G_{24} stand for merit consideration as it recorded higher grain yield in all the three environments as well as in pooled analysis.

Genetic diversity

D^2 analysis of the 25 genotypes confirmed the presence of high diversity among the types of their resolution into five clusters in E_1; nine clusters in E_2; eight clusters in E_3 and nine clusters in pooled analysis. Genotypes of different eco-geographic origin were grouped in a single cluster as well as in different clusters.

In E_1, cluster I consisted of three genotypes originating from Tamil Nadu (Aduthurai, Bhavani and Coimbatore). In E_2, cluster I comprised of two genotypes originating from Tamil Nadu and Andhra Pradesh. Cluster I in E_3, consisted of two genotypes both originating from Tamil Nadu. In pooled analysis cluster I comprised of three genotypes, all originating from Tamil Nadu (Table 1, 2, 3, and 4).

Interestingly, it was found out that the genotypes originating from Tamil Nadu, especially Aduthurai were found scattered in different clusters in the different environments and pooled analysis, those from Aduthurai may be from different sources. Out of the 12 genotypes of Aduthurai origin, one genotype was grouped in cluster I, one genotype was grouped in cluster II, eight genotypes were grouped in cluster III and yet other two genotypes were grouped in cluster IV in E_1. In E_2, out of the 12 genotypes of Aduthurai, two genotypes were grouped in cluster II, another two genotypes were grouped in cluster IV, another six genotypes were grouped in cluster V and yet another two genotypes were grouped in cluster VI. In E_3, out of the 12 genotypes of Aduthurai, three genotypes were grouped in cluster II, one genotypes was grouped in cluster III; one genotype was grouped in cluster V and yet another seven genotypes were grouped in cluster VI. In pooled analysis out of the 12 genotypes of Aduthurai origin, three genotypes were grouped in cluster I, one genotype was grouped in cluster II, six genotypes were grouped in cluster V and yet another two genotypes were grouped in cluster VII. Such a similar pattern of distribution was also found with the genotypes of Coimbatore and IRRI (Philippines) origin. This indicated the presence of wide range of genetic diversity among the genotypes of the present study. These genotypes could well be exploited in recombination breeding.

Crosses among the divergent parents are likely to yield desirable combinations. Therefore, a crossing programme should be initiated between the genotypes belonging to different clusters. In this context, two important points to be considered are : (i) choice of the particular cluster from which genotypes are to be used as parent in crossing programme and (ii) selection of particular genotype from selected groups.

The greater the distance between two clusters, the wider the genetic diversity among the parents to be included in hybridization programme. Parents combining high yield potential with wider genetic diversity are likely to yield superior segregants within in short period.

The tendency of genotypes from diverse geographic regions to group together in one cluster or scattered distribution of genotypes of same origin in different clusters has been observed in the present study. The studies of several workers confirmed this tendency, that the geographical diversity could not be directly related in any of the crops, tobacco (Murthy, 1965); linseed, wheat, sorghum (Murthy and Arunachalam, 1966); cotton (Singh and Gupta, 1968); safflower (Ranga Rao *et al.*, 1980); sesame (Thirugnana Kumar, 1991); rice (Singh, 1983; Ratho, 1984; Pande and Ghora, 1987; Singh *et al.*, 1987; De and Rao, 1987; Roy and Panwar, 1993; Vivekanandan and Sukanyasubramaniyan, 1993; Sarawgi and Shrivastava, 1996; Sawant *et al.*, 1996; Rao and Gomathinayagam, 1997; Bansal *et al.*, 1999; Singh *et al.*, 1999; Kandhola and Panwar, 1999; Karthikeyan, 2002; Venkatesan, 2004 and Sree Krishna, 2005).

Genetic drift and selection in different environments could cause greater diversity than geographical distance (Murthy, 1965). However geographical isolation could be important in the alteration of the breeding structure as pointed out by Wallace (1963).

Cluster V in E_1; cluster IX in E_2, cluster VII and VIII in E_3 and clusters VIII and IX recorded the lowest intra cluster distance value, cluster II in E_1; cluster I in E_2 and E_3 and cluster II in pooled analysis showed lower intra-cluster distance values. This may indicate the closeness of the genotypes. This may be explained on the basis that yield being a complex character of polygenic inheritance, similar phenotypes could be produced by many different combinations of genes and such combinations may have similar selective advantage (Singh and Gupta, 1968). Further, under constant selection on the segregating populations such similar types are expected to be established (Table 5, 6, 7 and 8).

The intra-cluster distance was maximum in cluster IV in E_1. It included as many as 11 genotypes. Intra-cluster distance was maximum in cluster VIII in E_2. It included three genotypes. Intra-cluster distance was maximum in cluster VI in E_3. It included as many as 11 genotypes. Intra-cluster distance was maximum in cluster VII of pooled analysis. It included as many as six genotypes. This may indicate that these genotypes were dissimilar. The limited gene exchange between these types or selection for diverse characters could be responsible for such high intra-cluster divergence.

In E_1, inter-cluster distance was maximum between the clusters III and V. Cluster III had lowest mean for days to 50 per cent flowering and highest mean for number of grains per panicle. The genotypes originating from these clusters may be recombined to evolve early maturing but high yielding lines. In E_2, inter cluster distance was maximum between the clusters V and IX. Cluster V had lowest mean for days to 50 per cent flowering as well as plant height at maturity. Cluster IX had high mean for 100 grain weight. The genotypes of these clusters may be combined for evolving early maturing dwarf genotypes with yield potential. In E_3, the inter-cluster distance was maximum between clusters II and VII, cluster II recorded lowest mean for days to 50 per cent flowering as well as plant height at maturity. Cluster VII had high mean for number of productive tillers. The genotypes originating from these clusters may be recombined to evolve early maturing but high yielding lines. In pooled analysis, the inter-cluster distance was maximum between cluster V and IX. Cluster V recorded lowest mean for days to 50 per cent flowering and highest mean for number of grains per panicle. The genotype of these clusters may be crossed to develop early maturing but high yielding lines (Table 9, 10, 11 and 12).

As pointed out by Sokal (1965), the choice of characters is important in multivariate analysis and the choice made in the present study appears to be appropriate. Cyclic breeding and crossing may be helpful in bringing new genes into a population of rice and thus, expanding the range of adaptation. As brought out by Mather (1943), polygenic variability, which is necessary for prospective, adaptive and evolutionary change need not exists as true phenotypic variation which effect fitness. Therefore, phenotypic uniformity with genetic diversity within the population appears to be very useful in rice. In fact, this is possible with a proper use of the suggested parents in the formation of gene complexes to replace the existing varieties.

It can be concluded that there is no relationship between genetic and geographical diversity in the material under investigation. Inter and intra-cluster distances were also not related. The grouping pattern was not consistent across environments, indicating that genetic diversity do depend on the environment and that genotype ´ environment affect genetic diversity.

It is highly desirable that the extent of genetic divergence between the populations reflected through any analysis, should be comparatively stable over seasons and environments, to be of utility to plant breeders. The differential grouping of different genotypes of varying origin in different environment does not mean that a breeder should study the divergence in one environment. It necessitated the need of studying the divergence in varying seasons as well as environments.

The approach suggested by Thirugnana Kumar (1991) is followed in the present study to suggest the parents based on the divergence pattern in varying environments.

a. One may select the parents on the basis of divergence exhibited in the richest and most productive environment, because it provides opportunity for the fullest expression of genetic potential of a genotype. Grain yield was higher in E_1. Using this criterion, divergent genotypes in E_1 may be selected from different clusters having high inter-cluster distances. Clusters III and V showed maximum inter-cluster distance in E_1. Cluster III composed of eight genotypes and cluster V composed of only one genotype. Cluster III recorded low mean for days to 50 per cent flowering and high mean for number of filled grains per panicle. Cluster V registered high mean for 100 grain weight. Hence by effecting crosses among the genotypes gathered in these clusters, one may expect earliness coupled with higher grain yield.
b. Selection of parents can also be made on the basis of divergence which is consistent over three environments. This can be taken as reliable indication of genetic divergence. Utilizing this criterion among the genotypes studied, the genotype namely Bhavani was gathered in cluster I in all the three environments (Table 4). Hence, it may be suggested that one may effect crosses between Bhavani and 21 out of 25 genotypes studied (*i.e.* except ADT 36, BPT 5204 and CO 47 which were gathered in cluster I). It may result in high yielding segregants (Table).
c. One may argue that the divergence expressed in pooled analysis may be reliable estimate and therefore should be used for selecting the parents. If this criterion is followed, the genotypes that were grouped in nine clusters may be crossed to evolve heterotic lines and segregants.

Out of the three approaches, the approach 'b' may be preferred, as it suggested the genotypes based on consistency in the divergence pattern. The approach 'a' suffers from environmental influence and approach 'c' suffers from the under estimation of genetic divergence because measures of divergence estimated in different environments may cancel each other in the pooled analysis.

Table 1. Composition of D^2 clusters – E_1

Clusters	Number of genotypes	Name of genotypes
I	3	ADT 36, Bhavani, CO 47
II	2	ADT 47, ASD 19
III	8	ADT 37, ADT 38, ADT 39, ADT 40, ADT 41, ADT 42, ADT 43, ADT 44
IV	11	ADT 45, ADT 48, BPT 5204, CO 43, CO 45, CR I009, IR 36, IR 64, Pusa Basmathi, TKM 9, TRY 1
V	1	White Ponni

Table 2. Composition of D^2 clusters – E_2

Clusters	Number of genotypes	Name of genotypes
I	2	Bhavani, BPT 5204
II	2	ADT 43, ADT 47
III	2	CO 47, ADT 47
IV	2	ADT 39, ADT 41
V	6	ADT 36, ADT 37, ADT 38, ADT 40, ADT 42, ADT 44
VI	5	ADT 45, ADT 48, ASD 19, CO 43, CO 45
VII	2	IR 64, TKM 9
VIII	3	IR 36, Pusa Basmathi, TRY 1
IX	1	White Ponni

Table 3. Composition of D^2 clusters – E_3

Clusters	Number of genotypes	Name of genotypes
I	2	Bhavani, CO 47
II	3	ADT 36, ADT 42, ADT 44
III	2	ADT 47, ASD 19
IV	2	IR 64, TKM 9
V	3	ADT 37, BPT 5204, CR 1009
VI	11	ADT 38, ADT 39, ADT 40, ADT 41, ADT 43, ADT 45, ADT 48, CO 43, C0 45, IR 36, Pusa Basmathi
VII	1	TRY 1
VIII	1	White Ponni

Table 4. Composition of D^2 clusters - pooled analysis

Clusters	Number of genotypes	Name of genotypes
I	3	ADT 36, ADT 42, ADT 43
II	2	ADT 47, ASD 19
III	2	CO 47, CR 1009
IV	2	Bhavani, BPT 5204
V	6	ADT 37, ADT 38, ADT 39, ADT 40, ADT 41, ADT 44
VI	2	IR 64, TKM 9
VII	6	ADT 45, ADT 48, CO 43, CO 45, IR 36, Pusa Basmathi
VIII	1	TRY 1
IX	1	White Ponni

Table 5. Inter and intra (diagonal) cluster average of D^2 and D (values in parentheses) and the extent of diversity among the clusters - E_1

Clusters	I	II	III	IV	V
I	142.41 (11.93)	116.16 (10.77)	125.95 (11.22)	181.77 (13.48)	524.18 (22.89)
II		5.22 (2.28)	158.63 (12.59)	198.46 (14.08)	292.78 (17.11)
III			92.86 (9.63)	279.87 (16.72)	732.28 (27.06)
IV				142.75 (11.94)	365.27 (19.11)
V					0.00 (0.00)

Table 6. Inter and intra (diagonal) cluster average of D^2 and D (values in parentheses) and the extent of diversity among the clusters - E_2

Clusters	I	II	III	IV	V	VI	VII	VIII	IX
I	4.43 (2.10)	25.98 (5.09)	19.65 (4.43)	95.11 (9.75)	195.07 (13.96)	88.97 (9.43)	106.31 (10.31)	354.29 (18.82)	396.93 (19.92)
II		12.94 (3.59)	62.71 (7.91)	74.21 (8.61)	132.07 (11.49)	113.43 (10.65)	178.89 (13.37)	496.86 (22.29)	524.34 (22.89)
III			13.20 (3.64)	153.33 (12.38)	285.35 (16.89)	88.00 (9 38)	128.51 (11.33)	265.17 (16.28)	298.25 (17.27)
IV				14.76 (3.84)	152.52 (12.35)	216.58 (14.71)	324.72 (18.02)	519.81 (22.79)	739.86 (27.20)
V					188.31 (13.72)	314.07 (17.72)	362.75 (19.04)	860.52 (29.33)	928.47 (30.47)
VI						180.14 (13.42)	208.00 (14.42)	427.06 (20.66)	368.18 (19.18)
VII							41.97 (6.47)	476.68 (21.83)	454.21 (21.31)
VIII								193.00 (13.89)	389.31 (19.73)
IX									0.00 (0.00)

Table 7. Inter and intra (diagonal) cluster average of D^2 and D (values in parentheses) and the extent of diversity among the clusters - E_3

Clusters	I	II	III	IV	V	VI	VII	VIII
I	3.33 (1.82)	121.31 (11.01)	46.79 (6.84)	177.34 (13.31)	95.34 (9.76)	83.09 (9.11)	266.00 (16.31)	137.48 (11.72)
II		120.73 (10.98)	107.34 (10.36)	342.28 (18.50)	181.71 (13.48)	149.19 (12.21)	567.73 (23.82)	401.16 (20.02)
III			13.38 (3.65	367.02 (19.15)	177.69 (13.33)	111.40 (10.55)	488.81 (22.10)	210.56 (14.51)
IV				27.28 (5.22)	184.08 (13.56)	272.42 (16.50)	124.98 (11.18)	267.44 (16.35)
V					135.51 (11.64)	155.96 (12.48)	274.23 (16.56)	286.58 (16.92)
VI						158.12 (12.57)	389.90 (19.74)	273.66 (16.54)
VII							0.00 (0.00)	252.35 (15.88)
VIII								0.00 (0.00)

Table 8. Inter and intra (diagonal) cluster average of D^2 and D (values in parentheses) and the extent of diversity among the clusters – pooled analysis

Clusters	I	II	III	IV	V	VI	VII	VIII	IX
I	212.90 (14.59)	236.70 (15.38)	577.47 (24.03)	381.60 (19.53)	245.09 (15.65)	1354.01 (36.79)	765.85 (27.67)	1931.34 (43.94)	2054.24 (45.43)
II		18.62 (4.31)	213.38 (14.60)	126.25 (11.23)	362.46 (19.03)	900.69 (30.01)	326.62 (18.07)	1229.51 (35.06)	1077.60 (32.82)
III			24.33 (4.93)	50.18 (7.08)	589.17 (24.27)	309.21 (17.58)	169.70 (13.02)	472.68 (21.74)	773.37 (27.81)
IV				29.67 (5.44)	369.45 (19.22)	472.43 (21.73)	198.23 (14.08)	698.58 (26.43)	1009.32 (31.77)
V					220.18 (14.83)	1321.43 (36.35)	763.57 (27.63)	1768.58 (42.05)	2241.63 (47.34)
VI						104.56 (10.22)	552.13 (23.49)	276.82 (16.63)	935.55 (30.58)
VII							328.71 (18.13)	614.69 (24.79)	700.83 (27.76)
VIII								0.00 (0.00)	722.05 (26.87)
IX									0.00 (0.00)

Table 9. Cluster means - E_1

Clusters	Days to 50 per cent flowering (days)	Plant height at maturity (cm)	Total number of tillers per plant	Number of productive tillers per plant	Number of filled grains per panicle	100 grain weight (g)	Grain yield per plant (g)
I	91.66	86.26	24.46	19.11	137.13	1.75	28.30
II	92.83	83.29	22.86	17.26	134.20	2.25	26.33
III	81.87	92.16	23.05	18.15	164.52	1.75	27.11
IV	107.24	103.65	25.39	19.86	140.25	1.71	27.95
V	121.00	110.20	16.86	13.60	127.46	2.56	23.43

Table 10. Cluster means - E_2

Clusters	Days to 50 per cent flowering (days)	Plant height at maturity (cm)	Total number of tillers per plant	Number of productive tillers per plant	Number of filled grains per panicle	100 grain weight (g)	Grain yield per plant (g)
I	101.66	94.11	23.33	18.46	138.13	1.72	26.22
II	93.16	90.17	24.00	19.03	130.40	1.92	27.12
III	106.16	97.75	23.96	19.13	124.78	1.87	28.02
IV	84.66	94.37	22.53	18.16	180.13	1.58	26.08
V	82.44	83.15	22.71	17.92	148.88	1.69	24.96
VI	103.93	93.59	23.58	18.64	134.90	1.98	28.79
VII	116.50	89.63	21.76	17.25	124.23	1.21	22.37
VIII	119.66	118.55	24.23	19.24	131.28	1.67	27.24
IX	127.33	108.73	15.80	12.46	124.26	2.50	19.06

Table 11. Cluster means - E_3

Clusters	Days to 50 per cent flowering (days)	Plant height at maturity (cm)	Total number of tillers per plant	Number of productive tillers per plant	Number of filled grains per panicle	100 grain weight (g)	Grain yield per plant (g)
I	107.33	91.30	22.80	18.43	127.70	1.83	27.02
II	90.22	77.83	20.64	16.32	131.60	1.76	22.73
III	101.16	87.75	21.46	17.60	130.96	2.19	25.55
IV	118.33	84.51	21.53	17.48	132.56	1.1O	21.97
V	99.44	94.85	22.38	18.26	159.94	1.48	25.48
VI	100.81	89.20	22.11	17.76	129.50	1.79	25.12
VII	128.66	111.84	22.43	18.46	126.46	1.25	20.98
VIII	128.66	99.63	14.46	11.60	114.33	2.20	18.38

Table 12. Cluster means of pooled analysis

Clusters	Days to 50 per cent flowering (days)	Plant height at maturity (cm)	Total number of tillers per plant	Number of productive tillers per plant	Number of filled grains per panicle	100 grain weight (g)	Grain yield per plant (g)
I	84.70	80.57	21.54	16.81	133.20	1.88	23.52
II	96.44	89.74	22.53	17.67	130.36	2.16	25.97
III	106.22	96.63	24.52	19.41	124.15	1.72	27.57
IV	100.94	94.84	23.95	19.06	140.99	1.74	27.21
V	84.50	88.52	22.43	17.96	160.75	1.62	26.26
VI	115.72	89.09	22.52	18.07	132.78	1.19	23.02
VII	107.40	101.51	24.35	19.26	136.10	1.90	28.31
VIII	123.22	116.07	23.37	18.66	130.66	1.31	23.84
IX	126.00	106.19	15.71	12.55	122.02	2.42	20.29

3

Genetic Diversity in Sesame [*Sesamum indicum* (L.)]

Thirugnanakumar, S. and A. Anandan

In recent years, we are just experiencing marginal surplus production in cereals leading to self-sufficiency through green revolution. However, the shortage of production in edible oils has drawn the attention for the increased cultivation of oil seeds, and increased productivity per unit area to provide sustainable food and nutrition security (Swaminathan, 1989).

The importance of oil seeds cannot be over emphasized in the Indian economy. Oilseed occupied nearly 19 million hectares, next only to food grains, in 1986-1987. Yet, the domestic production in the country has lagged far behind the increasing demand for edible oils, necessitating a massive import of edible oils which lead to the drain of the valuable foreign exchange. The total import bill on edible oil has increased from 1.41 per cent in 1970-71 to 3.05 per cent in 1986-87. Even then, the annual per capita availability of edible oils registered only a nominal increased from 3.5 kg to 5.9 kg during the same period, against the nutritional norm of 14 kg (Suresh Pal, 1989). Due to the concerted efforts made through the 'Technology Mission on Oilseeds', oil seeds production reached a peak of 178.9 lakh tonnes during 1988-'89 in India (GOI, 1990a). This is an impressive achievement in Indian Agriculture, to reduce the bill on the import of edible oils from 4.11 per cent (1987-'88) to 2.58 per cent (1988-'89) (Gill, 1990). Further efforts in increasing oilseeds production definitely will go a long way in drastically reducing the present import bill.

Among the oilseed crop, sesame (*Sesamum indicum* L.), "The Queen of Oil Seeds", is one of the important ancient and traditional crops occupying third position in terms of acreage on the oilseeds map of the country. It is commonly known as Gingelly, Til, Benniseed and Simsim. It is an important source of edible oil for the nutritional refugees of most part of the world, especially for India. The oil is rich in oleic and linoleic acid. In addition, it serves as a good source of proteins, vitamins and minerals (Nagaraj, 1990). Its high keeping quality deserves special mention.

Sesame is being raised in India in an area of 24.33 lakh hectares with an average productivity of 274 kg per hectare (GOI, 1990). In Tamil Nadu, this crop occupies an area of 1.21 lakh hectares with an average productivity of 298 kg per hectares (GOI, 1990b). This is far below the recorded high yield of 2000 kg per hectare in Yugaslovia (FAO, 1986). A further increase in sesame productivity needs intensive research in plant breeding and genetics. So far, sesame breeding research was carried out with much emphasis on seed yield and its component characters. Further more, most of the studies in a single season and the application of different biometrical techniques with a single set of data has not been much effective in rewarding in the practical plant breeding.

Genetic divergence among parents is important in self pollinated crops because a cross involving genetically diverse parents is likely to produce high heterozygous as well as heterotic effect. In such a cross, more variability could be expected in the segregating generations due to accumulation of the genes of both the parents. It is known that estimates of genetic divergence are also influenced by environments and seasons. Therefore, if the divergence analysis is conducted in a number of seasons and parents are selected on the basis of consistent divergence over the seasons, the problem of influence of seasons on divergence analysis can be largely overcome.

Genetic divergence, as one of the criteria of selection of parents, is considered in plant breeding as early as 1962 (Murty *et al.*, 1962; Timothy, 1963). Later Joshi and Dhawan, (1966), Murthy and Arunachalam (1966), Arunachalam *et al.* (1984) and Lefort - Buson (1987) also stated that for exploiting heterosis as a means of increasing production it is necessary to have parents of maximum genetic divergence. The more diverse the parents, more is the chance for pronounced heterotic effects and increased spectrum of variability in the segregating generations.

Limited information on genetic diversity in sesame is available so far (Table 13). Thirugnanakumar (1991) made a study to collect information on genetic diversity in 60 genotypes of different eco-geographic origin. He studied genetic diversity in three different seasons, for 16 traits, analysed individually as well as over three seasons. The relative contributions of the sixteen characters to the total diversity were also brought out. Mahalanobis D^2 statistic has been utilized as a powerful tool in quantifying the degree of divergence at genotypic level (Trehan *et al.*, 1974; Govinda Rasu, 1982; Rathinaswamy and Jagathesan, 1984; Thangavelu and Rajasekaran, 1983; Dhamu *et al.*, 1983 and Reddy, 1986; John Joel, 1987).

D^2 analysis of the sixty genotypes confirmed the presence of high diversity among the types by their resolution into twelve clusters in Season I, six clusters in Season II and pooled analysis, and eight clusters in Season III. Genotypes of different eco-geographic origin were grouped in a single cluster as well as in different clusters (Tables 14, 15, 16 and 17).

Cluster in SI consisted sixteen genotypes originating from Tamil Nadu (eight genotypes), Gujarat (two genotypes), Israel (one genotype), Greece (one genotype) and four other genotypes for which the origin is not traceable. In SII, a total of fifty one genotypes representing Tamil Nadu (26 genotypes), Andhra Pradesh (Five genotypes) Gujarat (two genotypes), Maharashtra (three genotypes), and one genotype from each of Rajasthan, Ceylon, Japan, Tanzania, Venezuela, Israel, Greece and eight genotypes of non-traceable origin were clustered in cluster I. Cluster I in S III also showed fifty one genotypes inclusive of twenty three genotypes of Tamil Nadu, five genotypes of Andhra Pradesh, four genotypes of Gujarat, two genotypes of Maharashtra, and one genotype from each of Punjab, Haryana, Pakistan, Tanzania, Venezuela, Israel, Greece and ten genotypes of non-traceable origin. But in pooled analysis only forty seven genotypes were grouped in cluster I. It consisted twenty four genotypes of Tamil Nadu, five genotypes of Andhra Pradesh, three genotypes of Gujarat, two genotypes of Maharashtra, one genotype from each of Pakistan, Tanzania, Venezuela, Israel and Greece. It also had eight genotypes of non-traceable origin. The other clusters also consisted different genotypes of varying origins (Table 18).

On the other hand, the twenty eight genotypes originated from Tamil Nadu were found scattered in different clusters in the different seasons and pooled analysis. Those from Tamil Nadu might also be from different sources. In SI, out of that 28 genotypes, eight genotypes were found in cluster I, fourteen in cluster II, one genotype in each of clusters IV, VI, VII, VIII, IX, X and XII. In S II, 26 were in cluster I and one in each of clusters II and IV were observed. In SIII, 23 in cluster I, two in cluster II and one in each of the clusters V, VI and VIII were recorded. Similarly, in the pooled analysis, 24 in cluster I, two in cluster II, and one in each of the clusters III and V were found. Such a similar pattern of distribution was also found with the genotypes of Gujarat and Maharashtra. This indicated the presence of wide range of genetic diversity among the types of Tamil Nadu, Gujarat and Maharashtra.

Table 13. Studies on genetic divergence in sesame

S. No.	No. of genotypes studied	No. of characters observed	No. of clusters observed	Maximum contributing characters	References
1	52	7	17	Plant height, days to maturity, seed number and capsules number	Trehen *et al.* (1974)
2	67	12	8	Days to maturity, days to flowering, seed yield and number of secondaries	Govinda Rao (1982)
3	9	5	4	Plant height and capsules per plant	Rathinaswamy (1980)
4	40	6	6	Oil content and days to maturity	Thangavelu and Rajaseakran (1983)
5	35	6	8	Total number of capsules per plant	Dhamu *et al.* (1983)
6	50	15	8	Plant height, capsules on branches, leaf area index, number of secondaries and days to maturity	Umesh Kumar Reddy (1986)
7	35	6	8	-	Ayyasamy *et al.* (1987)
8	9	17	5	1000 seed weight, seed yield and days to maturity	John Joel (1987)
9	64	20	15	Number of seeds per capsule, leaf area index, fruiting nodes, total capsules per plant, days to flowering and DMP	Anitha (1988)

Table 14. Composition of D^2 clusters – Season – I

Cluster number	Total No. of Genotypes	Genotypes	Cluster number	Total No. of Genotypes	Genotypes
Cluster – I	16	TMV.3			S. 0327
		B(1) S. 100043			S. 0516
		G. 51.266			S. 0604
		N. 369.50			S. 0639
		PK. 3			Si. 953
		S. 0431			Si. 1702
		Si. 0607			Si. 2678
		Si. 1249			C. 16
		Si. 1755			Co.1
		SI. 3175			IS. 299
		TMV. 4			NAL. 79.111.4.20
		A.T. 9			S. 0321
		IS. 102			S. 0325
		S. 0210			TMV. 6
		S. 0328	Cluster-III	3	Cianno 27/1
		S. 00646			NS. 78.221.32.1
Cluster-II	28	S. 00028			672-3-130-12
		S. 0218	Cluster-IV	2	IS. 298
		S. 0476			S. 00033
		Si. 930			
		Si. 1033	Cluster-V	2	Si. 1669
		Si. 1075	Cluster-VI	2	HT. 6
		Si. 1088			S. 00027
		Si. 1150	Cluster-VII	1	S. 00027
		Si. 1210	Cluster-VIII	1	Si. 1147
		Si. 1278	Cluster-IX	1	Si. 1160
		Si. 1703	Cluster-X	1	Is. 212
		Si. 2293			
		Marco. 7	Cluster-XI	2	EC. 137935
					IS. 232
			Cluster-XII	1	S. 0551

Table 15. Composition of D^2 clusters – Season – II

Cluster number	Total No. of Genotypes	Genotypes	Cluster number	Total No. of Genotypes	Genotypes
Cluster – I	51	IS. 298			Si. 953
		S. 00027			Si. 1249
		S. 00028			Si. 1702
		S. 00033			Si. 1755
		S. 0218			Si. 2678
		Si. 0476			Si. 3175
		Si. 030			TMV. 4
		Si. 1033			C. 16
		Si. 1075			EC. 137935
		Si. 1147			IS. 102
		Si. 1150			IS. 232
		Si. 1160			IS. 299
		Si. 1209			NAL. 79.111.4.20
		Si. 1210			S. 0210
		Si. 1278			S. 0321
		Si. 1669			S. 0325
		Si. 1703			S. 0328
		Si. 2293			S. 0551
		TMV. 3			S. 00646
		B(1) S. 100043			TMV. 6

		G. 5S. 266			672-3-13-12
		JLT. 7	Cluster-II	5	Si. 108
		Margo 7			IS. 212
		PK. 8			A.T. 9
		S. 0327			Cannoo 27/1
		S. 0431			S. 0554
		S. 0516	Cluster-III	1	N. 369.50
		S. 0604	Cluster-IV	1	CO. 1
		S. 0607	Cluster-V	1	HT. 6
		S. 0639	Cluster-VI	1	?

Table 16. Composition of D^2 clusters – Season – III

Cluster number	Total No. of Genotypes	Genotypes	Cluster number	Total No. of Genotypes	Genotypes
Cluster – I	51	IS. 298			Si. 1249
		S. 00027			Si. 1702
		S. 00028			Si. 1755
		S. 00033			Si. 2678
		S. 0218			Si. 3175
		Si. 0476			TMV. 4
		Si. 030			A.T. 9
		Si. 1033			C. 16
		Si. 1075			Cannoo 27/1
		Si. 1147			CO. 1
		Si. 1150			HT. 6
		Si. 1160			IS. 102
		Si. 1209			JS. 299
		Si. 1210			NAL. 79.111.4.20
		Si. 1278			NS. 78.221.32.1
		Si. 1669			S. 0210
		Si. 1703			S. 0321
		Si. 2293			S. 0328
		TMV. 3			S. 00646

		B(1) S. 100043			TMV. 6
		G. 5S. 266			672-3-13-12
		JLT. 7	Cluster-II	3	IS. 232
		Margo 7			S. 0551
		PK. 8			S. 0554
		S. 0327	Cluster-III	1	Si. 1209
		S. 0431	Cluster-IV	1	Si. 1669
		S. 0516	Cluster-V	1	TMV. 3
		S. 0607	Cluster-VI	1	S. 0604
		S. 0639	Cluster-VII	1	EC. 137935
		Si. 953	Cluster-VIII	1	S. 0325

Table 17. Composition of D^2 clusters – Pooled

Cluster number	Total No. of Genotypes	Genotypes	Cluster number	Total No. of Genotypes	Genotypes
Cluster – I	47	IS. 298			Si. 1755
		S. 00027			Si. 2678
		S. 00028			Si. 3175
		S. 00033			TMV. 4
		S. 0218			C. 16
		Si. 0476			Cannoo 27/1
		Si. 930			Co. 1
		Si. 1033			IS. 102
		Si. 1075			IS. 299
		Si. 1088			NAL. 79.111.4.20
		Si. 1147			S. 0210
		Si. 1150			S. 0321
		Si. 1210			S. 0325
		Si. 1278			S. 0328
		Si. 1703			S. 00646
		Si. 2293			TMV. 6
		B(1) S. 100043			672-3-130-12
		JLT. 7	Cluster-II	5	Si. 1160
		Margo. 7			Si. 1209
		N. 369.50			TMV. 3
		PDK. 3			A.T. 9

	S. 0327			HT. 6
	S. 0431	Cluster-III	3	IS. 212
	S. 0516			NS. 78.221.32.1
	S. 0604			S. 0554
	S. 0607	Cluster-IV	2	EC. 137935
	S. 0639			IS. 232
	Si. 953	Cluster-V	2	G. 51.266
	Si. 1249			S. 0551
	Si. 1702	Cluster-VI	1	Si. 1669

Table 18. Overall clustering pattern

Genotypes and origin	Seasons			
	SI	SII	SIII	Pooled analysis
Tamil Nadu				
G2	VII	I	I	I
G3	II	I	I	I
G4	IV	I	I	I
G5	II	I	I	I
G6	II	I	I	I
G11	VIII	I	I	I
G12	II	I	I	I
G13	IX	I	I	II
G15	II	I	I	I
G20	I	I	V	II
G27	I	I	I	I
G28	II	I	I	I
G29	I	I	I	I
G30	II	I	I	I
G31	II	I	VI	I
G32	I	I	I	I
G33	II	I	I	I
G34	II	I	I	I
G40	I	I	I	I
G44	II	IV	I	I
G52	I	I	I	I
G53	II	I	I	I
G54	II	I	VIII	I
G55	I	I	I	I
G56	XII	I	II	V
G57	VI	II	II	III
G58	I	I	I	I
G59	II	I	I	I
Andhra Pradesh				
G7	II	I	I	I
G8	II	I	I	I
G18	II	I	I	I
G36	II	I	I	I

G42	II	I	I	I
Gujarat				
G1	IV	I	I	I
G26	I	III	I	I
G41	I	II	I	II
G49	II	I	I	I
Maharashtra				
G16	II	I	I	I
G17	V	I	IV	VI
G24	II	I	I	I
Rajasthan				
G48	XI	I	II	IV
Punjab				
G23	X	II	I	III
Haryana				
G46	VI	V	I	II
Ceylon				
G14	V	I	III	II
Pakistan				
G10	III	II	I	I
Japan				
G45	XI	I	VII	IV
Venezuela				
G25	II	I	I	I
Greece				
G35	I	I	I	I
Non-traceable				
G9	II	I	I	I
G19	II	I	I	I
G21	I	I	I	I
G22	I	I	I	V
G38	II	I	I	I
G39	I	I	I	I
G43	III	II	I	I
G47	I	I	I	I
G51	III	VI	I	III
G60	III	I	I	I

Trehan *et al.* (1974) and Joel (1987) also contempted with a similar conclusion, that a wide genetic diversity exists among the genotypes of Tamil Nadu. In fact the observation of clusters carrying only one genotype that too of Tamil Nadu origin, confirmed this. These genotypes can be well exploited in heterosis breeding.

The tendency of genotypes from diverse geographic regions to group together in one cluster or scattered distribution of genotypes of same origin in different clusters has been observed in the present study. The studies of several workers confirmed this tendency, that the geographical distribution and genetic diversity could not be directly related in any of the crops, tobacco (Murty, 1965), linseed, wheat, sorghum (Murty and Arunachalam, 1966), cotton (Singh and Gupta, 1968), greengram (Singh *et al.*, 1976), safflower (Ranga Rao *et al.*, 1980), and sesame (Govinda Rasu, 1982; Thangavelu and Rajasekaran, 1983, Reddy, 1986 and John Joel, 1987). This is not surprising, since we know that there is free exchange of genetic material for breeding purposes throughout the world.

Genetic drift and selection in different environments could cause greater diversity than geographical distance (Murty *et al.*, 1965). However, geographical isolation could be important in the alteration of the breeding structure as pointed out by Wallace (1963).

In certain cases, the effect of geographic origin appeared to have influenced the clustering pattern. Five types collected from Andhra Pradesh were clustered in a single cluster only, in all the analysis. The only one genotype of Rajasthan (G 48) and one genotype of Japan (G 45) were mostly grouped in a single cluster. Probably, G 45 may be introduced from Japan to Rajasthan, where some natural infilteration of genes might have happened among these genotypes. Such an effect of geographic origin on clustering pattern has been recorded by Jeswani *et al.* (1970) in linseed, Ziauddin Ahamad *et al.* (1980) in triticale, Govinda Rasu (1982) and Joel (1987) in sesame. It suggests that although geographical distribution could not be taken as a sole criterian of genetic diversity in sesame, the influence of the former could still be traced.

Cluster III including three genotypes in SI, cluster I consisting fifty one genotypes in S II, cluster II representing three genotypes in S III and Cluster V having two genotypes in pooled analysis showed the least intracluster divergence in the concerned seasons and analysis. This indicated the closeness of types. This may be explained on the basis that yield being a complex character of polygenic inheritance similar phenotypes could be produced by many different combinations of genes and such combinations may have similar selective advantage (Singh and Gupta, 1968). Further, under constant selection on the segregating population such similar types are expected to be established. The highest intra-cluster divergence was recorded, for the cluster XI in SI, II in SII, I in S III and III in the pooled analysis. While the limited gene exchange between the types or selection for diverse characters could be responsible for such high intra-cluster divergence. The penetrance and expressivity of the genotypes have also to be considered in a given set of season, as the cluster and the composition, differed widely season after season.

Highest inter-cluster distance was observed between clusters VII and XI in S I. Cluster VII exhibited higher mean values for number of branches and seed density. Cluster XI expressed maximum mean value for 1000 seed volume. So by crossing the genotypes which were grouped in VII and XI clusters will be useful to improve the seed yield as well as 1000 seed weight. Number of branches and seed density were positively and significantly associated with seed yield and 1000 seed weight. Thousand seed volume was positively and significantly associated with 1000 seed weight, it is possible to improve both seed yield and seed size (Thirugnanakumar, 1991) (Table 19).

In S II, clusters III and VI exhibited highest inter-cluster distance. Cluster III showed maximum mean value for plant height at maturity number of branches, number of flowers, number of capsules, oil content and total dry matter production. Excepting oil content all the other traits were positively and significantly associated with seed yield and seed size. Oil content mostly showed positive significant association with all these component traits. Cluster VI expressed minimum value for days to maturity and plant height. So there is a possibility to obtain genotypes with high seed yield, seed size and oil content coupled with early maturing, medium statured plant types (Tables 20, 21, 22, 23 and 24).

Similarly, for most of the traits which were positively and significantly associated with seed yield and seed size, cluster VI evidenced maximum mean value in S III. In this season cluster VII had maximum mean value for seed breadth and 1000 seed volume. These two clusters expressed maximum intercluster distance. Hence, there is a possibility to get higher seed yield, seed size coupled with higher oil content by inter-crossing among the genotypes of these clusters (Table 25).

In the pooled analysis, cluster III evinced minimum number of days to maturity and cluster IV had maximum mean value for capsule volume, seed length, seed breadth, 1000 seed volume and oil content. These two clusters showed maximum inter-cluster distance. By inter-mating among the genotypes of these clusters there is a possibility to obtain higher seed size with higher oil content. As cluster III also showed lesser plant height there is a possibility to get medium statured genotypes (Table 26).

As pointed out by Sokal (1965), the choice of the characters is important in multivariate analysis and the choice made in the present study appears to be appropriate. Cyclic inbreeding and crossing may be helpful in bringing new genes into a population of sesame and thus, expanding the range of adoptation. As brought out by Mather (1943) polygenic variability which is necessary for prospective, adaptive and evolutionary change need not exist as true phenotypic variation which effect fitness. Therefore, phenotypic uniformity with genetic diversity within the population appears to be very useful in sesame. In fact, this is possible with a proper use of the suggested parents in the formation of gene complexes to replace the existing varieties.

It can be concluded that there is no relationship between genetic and geographical diversity in the material under investigation (Thirugnanakumar, 1991). Inter and intra-cluster distances were also not related. The grouping pattern was not consistent across seasons, indicating that genetic diversity do depend on the season and that genotype x season interaction affect genetic diversity. This is not unexpected because genetic diversity itself depend on the penetrance and expression of genes and there by the expression of traits in a given set of environment or season.

It is highly desirable that the extent of genetic divergence between populations reflected through any analysis, should be comparatively stable over seasons and environments to be of utility to plant breeders. The differential grouping of different genotypes of varying origin in different seasons, does not mean that a breeder should study divergence in only one season. In fact, it really suggests the need of studying the divergence in varying seasons.

The approaches suggested by Singh and Gupta (1984) and Singh *et al.* (1988) were followed in the present study to suggest the parents based on the divergence pattern in varying seasons.

a. One may select the parents on the basis of divergence exhibited in the richest and most productive season, because it provides opportunity for the fullest expression of genetic potential of a genotype. The relative ranking of the three seasons in the present study revealed that S III was best for days to maturity, number of flowers, 1000 seed weight, 1000 seed volume, seed yield and harvest index. Since this season exhibited higher mean performance for 1000 seed weight, seed yield and harvest index, seemed to be the best season. Using this criterion, divergent genotypes in season III may be selected from different clusters having high inter-cluster distances. Clusters VI and VII showed maximum inter-cluster distance in S III. Only one genotype in each of these two clusters viz., G 31 and G 45 were found respectively. Eventhough these genotypes were high yielding they were not stable. So, they are not suitable for breeding programme. Among the other combinations of clusters viz., IV and V, IV and VII, VII and VIII, II and VI, II and IV and I and VII, cluster I and VII having the inter-cluster distances slightly lower than the highest were selected which consisted both high yielding stable genotypes and high yielding responsive (not stable) genotypes. These cross combinations will involve G 45 (Japanese origin) with fifty one genotypes of Cluster I.

b. Selection of parents can also be made on the basis of divergence which is consistent over the seasons. This can be taken as reliable indication of genetic divergence. Utilizing this criterion among the genotypes studied, following genotypes showed consistent divergence in all the three seasons:

i. G56 was consistently divergent with the following genotypes: G10, G23, G26, G41, G44 and G46.

ii. G 45 and G 48 were consistently divergence from G 10, G 23, G 26, G 41, G 44 and G 46. Added to this G 45 was also consistently different from G 57.

iii. G 14 and G 17 were consistently divergent from G 10, G 23, G 26, G 41, G 44, G 46 and G 57.

iv. G 23 was consistently divergent from G 14, G 17, G 31, G 45 and G 54.

v. G 46 was consistently divergent from G 14, G 17, G 20, G 31, G 45, G 48, G 54 and G 56.

It is interesting to conceive that the divergent genotypes viz., G 14, G 17, G 45, G 48 and G 56 were also consistently divergent from the divergent genotypes G 23 and G 46. But the divergence among these two groups were not consistent. Thus they exhibited highly dissimilar response to the season due to genotype and genotype x season inter-action. On the contrary, the following genotypes viz., G 21, G 27, G 29, G 32, G 35, G 37, G 39, G 40, G 47, G 52, G 55 and G 58 were highly iso-responsive, and they were grouped in cluster I in all the analysis.

c. One may also argue that the divergence expressed in pooled analysis may be reliable estimate and therefore should be used for selecting parents. If this criterion is followed genotypes viz., G 13, G 14, G 20, G 41 and G 46 were divergent from the rest of the sixty genotypes investigated. G 23 and G 51 were divergence from all other fifty eight genotypes.

G 45 and G 48 were divergent from the remaining 58 genotypes. Similarly, G 22 and G 56 were divergent from the genotypes G 1 to G 21, G 23 to G 55 and G 55 to G 60. G 17 was also divergent from the other fifty nine genotypes of the study.

It is highly interesting to note that the genotypes which were described as consistently divergent in the second (b) approach based on consistency in divergence in all the three seasons, also divergent in the pooled analysis excepting a small difference. In the second approach it was described that G 14 was consistently divergent from G 41 and G 46, but in pooled analysis they were grouped together (Cluster II). Similarly the genotype G 46 which was classified as consistently divergent from G 14, and G 20 were grouped together in cluster II in the pooled analysis. Excepting this small discrepancy both the classifications brought out the same result. Such a small discrepancy was also evidenced in the divergence pattern among the two divergent groups described at the end of approach ' b '. In the approach ' b [1], it was found that the divergent genotypes G 14, G 17, G 45, G 48 and G 56 consistently divergent from the divergent genotypes G 23, and G 46. But in the pooled analysis, G 14 and G 46 were grouped together (Cluster II).

On an over all consideration, both the approaches viz., based on consistency over seasons and pooled analysis hold good in the identification of the divergent genotypes. This overruled the dispute that pooled analysis may under estimate divergence because measures of divergence estimated in different seasons may cancel each other in the pooled analysis. In fact, this was supported by the realized more or less similar maximum intra-cluster distances in the pooled analysis as well as in the three different seasons. This again indicated a more or less similar cluster constitution. However, the cancellation effect was evidenced from the reduction of inter-cluster distance in the pooled analysis than that of individual season wise analysis.

In the present study, all the three method of approaches yielded a similar result confirming all the three approaches are suitable for selecting the divergent genotypes. A sesame breeder can select any of the three approaches based on his requirement urgency and policy changes. The following genotypes and crossings are suggested for cross breeding programme (Table 27).

i) Using the first criteria, (approach 'a') it may be concluded from the above discussion that the following cross combinations involving high yielding stable and high yielding responsive (unpredictable) genotypes are suitable to be used in future breeding programmes.

Table 19. Inter and intra (diagonal) cluster average of D^2 and D (values with in parenthesis) and the extent of diversity among the clusters – S I

Clusters	I	II	III	IV	V	VI	VII	VIII	IX	X	XI	XII
I	**380.6419 (19.51)**	781.4598 (27.95)	867.1466 (29.45)	1404.5502 (37.48)	612.6682 (24.75)	727.0004 (26.96)	1337.2725 (36.57)	942.9822 (30.71)	692.1716 (26.31)	934.8907 (30.57)	3392.0179 (58.24)	1174.3039 (34.27)
II		**343.0669 (18.52)**	673.8246 (25.96)	617.6536 (24.85)	900.2755 (30.01)	1668.7203 (40.85)	767.6436 (27.71)	517.6618 (22.75)	1316.1411 (36.28)	958.3986 (30.96)	5418.5959 (73.61)	2641.6388 (51.40)
III			**329.7550 (18.16)**	1299.661 (36.05)	1440.1532 (37.95)	1246.8694 (35.31)	907.5542 (30.13)	789.5700 (28.10)	1940.910 (44.06)	663.8875 (25.77)	5713.7495 (75.59)	2691.0563 (51.88)
IV				**412.9400 (20.32)**	1474.1323 (38.39)	2378.8845 (48.77)	981.9744 (31.34)	831.0914 (28.33)	1869.9245 (43.24)	1728.1690 (41.57)	6389.0018 (79.93)	3587.937 (59.90)
V					**452.0500 (21.26)**	1547.6095 (39.34)	1533.157 (39.16)	972.8355 (31.39)	688.6914 (26.24)	1492.1115 (33.39)	3651.0005 (60.42)	1655.6265 (40.69)
VI						**448.9100 (21.19)**	2359.5745 (48.58)	1991.3140 (44.62)	1238.3625 (35.19)	1114.7512 (33.39)	2726.9398 (52.22)	863.3516 (29.38)
VII							**0.00 (0.00)**	560.1706 (23.67)	2381.0500 (48.80)	842.3632 (29.02)	6731.2010 (82.04)	3691.6710 (60.76)
VIII								**0.00 (0.00)**	1861.763 (43.15)	1131.22 (33.63)	6171.851 (78.56)	3224.116 (56.78)
IX									**0.00 (0.00)**	1696.2140 (41.19)	2243.6175 (47.37)	716.8431 (26.77)
X										**0.00 (0.00)**	5431.3120 (73.70)	2387.535 (48.86)
XI											**532.090 (23.07)**	858.0136 (29.29)
XII												**0.00 (0.00)**

(*Source*: Thirugnanakumar, 1991)

Table 20. Inter and intra (diagonal) cluster average of D^2 and D (values with in parenthesis) and the extent of diversity among the clusters – S II

Clusters	I	II	III	IV	V	VI
I	361.7365 (19.02)	745.6643 (27.31)	763.0353 (27.62)	287.3988 (16.95)	849.9277 (29.15)	1674.2882 (40.92)
II		433.1283 (20.81)	1592.9838 (39.91)	807.4367 (28.42)	808.2223 (28.43)	980.5618 (31.31)
III			0.00 (0.00)	540.8708 (23.26)	1478.797 (38.46)	2488.392 (49.88)
IV				0.00 (0.00)	854.9784 (29.24)	1906.0630 (43.66)
V					0.00 (0.00)	738.4385 (27.17)
VI						0.00 (0.00)

(*Source*: Thirugnanakumar, 1991)

Table 21. Inter and intra (diagonal) cluster average of D^2 and D (values with in parenthesis) and the extent of diversity among the clusters – S III

Clusters	I	II	III	IV	V	VI	VII	VIII
I	**618.3064 (24.87)**	1648.0955 (40.60)	1638.8169 (40.48)	2269.3861 (47.64)	1807.3192 (42.51)	1990.2905 (44.61)	2911.5923 (53.96)	956.9554 (30.93)
II		**549.94 (23.45)**	1827.1253 (42.74)	3628.555 (60.24)	1296.1407 (36.00)	3680.7313 (60.67)	917.441 (30.29)	2545.6483 (50.41)
III			**0.00 (0.00)**	1172.687 (34.24)	3008.639 (54.85)	3924.147 (62.64)	2938.749 (54.21)	1167.901 (34.17)
IV				**0.00 (0.00)**	4362.581 (66.05)	3711.188 (60.92)	4555.343 (67.49)	1773.729 (4212)
V					**0.00 (0.00)**	1434.842 (37.87)	1935.241 (43.99)	3273.55 (57.21)
VI						**0.00 (0.00)**	4657.567 (68.25)	2520.048 (50.20)
VII							**0.00 (0.00)**	3839.323 (61.96)
VIII								**0.00 (0.00)**

(*Source*: Thirugnanakumar, 1991)

Table 22. Inter and intra (diagonal) cluster average of D^2 and D (values with in parenthesis) and the extent of diversity among the clusters – Pooled

Clusters	I	II	III	IV	V	VI
I	**153.0453 (12.37)**	359.0404 (18.95)	527.9698 (22.98)	1436.7764 (37.90)	516.9902 (22.74)	752.7905 (27.44)
II		**194.6906 (13.95)**	498.8438 (22.33)	732.0309 (27.06)	190.7513 (13.81)	658.4696 (25.66)
III			**549.7900 (23.45**	1496.2470 (38.68)	582.9825 (24.15)	1264.645 (35.56)
IV				**104.0300 (10.20)**	384.5471 (19.61)	1477.9685 (38.44)
V					**88.1400 (9.39)**	681.5823 (26.11)
VI						**0.00 (0.00)**

(*Source*: Thirugnanakumar, 1991)

Table 23. Cluster means for sixteen characters in sesame – S I

Characters	I	II	III	IV	V	VI	VII	VIII	IX	X	XI	XII
X_1	107.09	106.32	103.67	106.28	115.25	**102.10**	107.20	106.70	106.10	105.40	106.40	107.20
X_2	79.78	80.02	**40.81**	92.30	99.68	75.90	78.08	94.04	83.25	46.09	80.44	85.01
X_3	8.69	9.22	**5.97**	13.75	10.68	9.95	20.80	9.45	8.00	9.65	8.93	6.40
X_4	74.93	79.75	25.43	105.28	88.43	52.33	67.10	112.90	38.00	**22.60**	68.93	55.35
X_5	63.79	68.79	**17.43**	77.00	77.45	42.65	49.40	109.60	29.50	18.40	59.65	45.90
X_6	0.33	0.32	0.30	0.42	0.48	0.30	**0.13**	0.28	0.38	0.22	0.43	0.40
X_7	54.65	54.81	57.82	57.52	58.28	53.56	**50.28**	52.40	**59.12**	53.80	57.65	55.40
X_8	3.03	3.15	3.11	3.36	3.20	**2.89**	3.00	2.94	**3.52**	3.12	3.22	3.11
X_9	1.77	1.82	1.74	1.94	1.86	1.79	1.76	1.77	1.87	**1.63**	1.82	1.70
X_{10}	2.77	2.97	**2.30**	3.36	3.34	2.46	3.49	3.59	3.32	3.10	2.45	2.46
X_{11}	8.28	5.49	5.19	4.98	8.30	9.32	**4.97**	5.93	9.84	5.97	**13.47**	11.05
X_{12}	0.34	0.55	0.46	0.68	0.41	0.27	0.70	0.61	0.34	0.52	**0.19**	0.22
X_{13}	43.09	44.48	42.72	44.00	43.45	37.25	38.70	45.70	45.40	**32.75**	45.15	41.20
X_{14}	4.13	4.60	3.20	4.08	4.44	**2.55**	4.26	9.31	3.04	4.21	4.14	2.85
X_{15}	21.18	20.72	**10.72**	23.60	29.77	17.42	18.35	26.09	13.59	14.04	23.54	18.12
X_{16}	19.22		28.15	17.38	**11.44**	16.02	22.76	35.58	22.47	29.91	27.26	15.83

Bold indicates minimum cluster values.(*Source*: Thirugnanakumar, 1991) Underline indicates maximum cluster values.

Table 24. Cluster means for sixteen characters in sesame – S II

Characters	Clusters					
	I	II	III	IV	V	VI
X_1	105.11	104.12	104.30	105.30	97.50	**94.10**
X_2	53.97	37.94	64.10	56.00	46.61	**32.66**
X_3	8.45	8.14	10.80	9.20	**6.80**	7.00
X_4	76.17	**56.86**	132.00	91.00	61.40	66.70
X_5	57.95	38.60	70.30	64.10	**22.90**	27.75
X_6	0.37	**0.31**	0.35	0.41	0.53	0.39
X_7	55.50	**53.34**	55.80	55.80	56.20	55.60
X_8	3.10	3.00	**2.81**	3.41	3.13	2.83
X_9	1.63	1.63	**1.53**	1.79	1.73	1.56
X_{10}	3.10	3.00	3.28	3.56	3.12	**2.46**
X_{11}	4.77	4.88	4.62	4.72	5.31	**3.65**
X_{12}	0.65	0.62	0.71	0.76	**0.59**	0.68
X_{13}	42.95	37.74	45.75	43.65	41.15	**31.15**
X_{14}	5.12	3.64	6.80	7.29	**2.42**	3.65
X_{15}	21.64	20.53	33.50	23.19	**17.11**	21.99
X_{16}	24.78	19.03	20.44	31.52	**14.23**	16.42

Bold indicates minimum cluster values. (*Source*: Thirugnanakumar, 1991)Underline indicates maximum cluster values.

Table 25. Cluster means for sixteen characters in sesame – S III

Characters	Clusters							
	I	II	III	IV	V	VI	VII	VIII
X_1	96.95	93.77	116.45	129.30	92.90	99.40	**92.00**	100.00
X_2	50.83	41.89	46.85	**35.70**	64.02	72.90	43.83	37.71
X_3	8.80	8.63	7.75	**7.30**	11.00	10.90	8.35	9.30
X_4	76.97	65.33	**54.70**	60.50	117.20	144.30	79.10	64.20
X_5	53.77	37.13	17.10	**11.90**	102.65	124.55	18.65	37.15
X_6	0.34	**0.20**	0.39	0.29	0.36	0.29	0.22	0.29
X_7	50.18	**48.47**	51.00	50.60	51.00	49.20	48.90	51.60
X_8	3.03	2.81	2.78	2.85	2.85	3.02	2.95	**2.68**
X_9	1.65	1.60	1.68	**1.48**	1.75	1.54	1.81	1.51
X_{10}	3.81	4.72	**2.18**	2.79	5.39	4.65	3.22	3.42
X_{11}	6.66	11.54	8.32	8.90	10.46	5.06	14.57	**3.79**
X_{12}	0.60	0.43	0.27	0.32	0.51	0.92	**0.22**	0.90
X_{13}	42.22	39.85	**32.70**	42.30	41.70	44.90	43.90	38.35
X_{14}	5.84	4.29	0.68	**0.62**	13.82	17.52	1.79	0.99
X_{15}	21.09	20.21	12.54	**7.48**	40.81	41.33	9.97	7.83
X_{16}	26.22	20.87	**5.59**	8.30	33.83	42.51	17.94	12.78

Bold indicates minimum cluster values. (*Source*: Thirugnanakumar, 1991) Underline indicates maximum cluster values.

Table 26. Cluster mean for 16 characters in sesame – pooled

Sl. No.	Characters	Clusters					
		I	II	III	IV	V	VI
1	X_1	102.79	103.38	**99.57**	100.35	103.85	119.70
2	X_2	61.66	59.11	**40.28**	54.43	61.38	58.63
3	X_3	8.95	8.12	**7.67**	8.25	8.85	10.60
4	X_4	78.30	73.80	**49.40**	68.80	69.60	64.90
5	X_5	60.90	47.34	42.90	43.70	48.75	**33.70**
6	X_6	0.32	0.38	0.28	0.74	**0.27**	0.37
7	X_7	54.04	54.20	**52.40**	54.05	53.50	54.30
8	X_8	3.08	3.06	**2.90**	3.13	3.07	3.01
9	X_9	1.70	1.73	**1.60**	1.75	1.66	1.64
10	X_{10}	3.31	3.17	3.08	**3.06**	3.24	3.15
11	X_{11}	**5.77**	7.74	5.83	10.88	8.80	7.60
12	X_{12}	0.60	0.45	0.56	**0.35**	0.41	0.45
13	X_{13}	43.12	41.14	**36.60**	44.15	41.20	42.90
14	X_{14}	5.32	3.93	3.84	4.41	4.31	**2.55**
15	X_{15}	21.31	21.82	18.60	21.46	18.59	**16.25**
16	X_{16}	24.58	17.32	21.64	22.30	26.48	**16.24**

Bold indicates minimum cluster values. (*Source*: Thirugnanakumar, 1991) Underline indicates maximum cluster values.

Table 27. Genotypes and crossing suggested based on genetic diversity

Approach 'a'	Approach 'b'			Approach 'c'		
G1,G2,G3,G4,G5,G6,G7,G8, G9,G10,G11,G12,G13,G14, G15,G16,G17,G18,G9,G20, G21,G22,G23,G24,G25,G26, G27,G28,G29,G30,G31,G32, G33,G34,G35,G36,G37,G38, G39,G40,G41,G42,G43,G44, G45,G46,G47,G48,G49,G50, G51,G52,G53,G54,G55,G56, G57,G58,G59,G60	G14, G17, G45, G48, G56	´	G10,G23,G26, G41,G44,G46	G17, G45, G48, G56	´	G10,G23,G26, G46,G41,G44
	G14 G17	´	G57	G14	´	G10, G23, G26, G44
				G14, G17	´	G57
	G23 G46	´	G57	G23 G46	´	G31, G54
	G45	´	G31, G54	G45	´	G57
	G46	´	G20	G46	´	G56
Common genotypes and crosses: G45 ´ G10, G23, G26, G41, G44, G46						

EC 137935 (G 45)	x	S. 00028	(G 3)
	x	Si. 1150	(G 12)
	x	PDK.8	(G 27)
	x	S.0516	(G 30)

Genotype, G 45 is Japanese origin and G 3, G 12, G 27 and G 30 are Tamil Nadu origin. As both geographical and genetic diversity exists among these genoytpes, these cross combinations involving these genotypes are expected to give high heterotic hybrids in F_1. generation and useful segregants in the later generations. Added to this, one can expect heterosis for yield as well as response to the seasons or environments, because the genotypes G 3, G 12, G 27 and G 30 were high yielding stable with average responsiveness and G 45 was high yielding responsive (unpredictable).

ii) Using first and second criterion, the following genotypes which are common in both the approaches (' a ' and 'b'), also suggested for future breeding programme.

Ec 137935 (G 45) x	Si. 1088	(G 10)
x	IS.212	(G 23)
x	N.36950	(G 26)
x	A.T.9	(G 41)
x	Co.1	(G 44)
x	HT.6	(G 46)

iii) The divergence of above crosses were also confirmed by the third criterion, i.e., on the basis of pooled analysis and hence the above six crosses can also be utilized in breeding programme. However, when we consider high seed yield coupled with high mean performance of the traits which were positively associated with seed yield and seed size, priority must be given to the following crosses:

EC 137935 (G 45)	x N.36950	(G 26)
	x Co.1 (G 44)	

These three genotypes showed high mean seed yield. Added to this, G 26 exhibited high per se performance for number of branches and flowers, and G 44 expressed high per se performance for number of flowers, capsules, seed density and harvest index. Eventhough the behaviour of these three genotypes can not be predicted with reference to seed yield, they were highly divergent. So heterosis can well be expected for these traits.

The foregoing discussions and suggestions are purely based on high genetic divergence coupled with stability and average responsiveness to get heterosis for both yield and responsiveness. This is possible because yielding ability and response to environmental changes are two independent attributes of a genotypes and are governed by separate sets of gene systems, (Finlay and Wilkinson, 1963; Bucio-Alanis *et al.*, 1969; Bains, 1976; Verma *et al.*, 1978; Singh, 1982; Singh and Gupta, 1985 and Singh, 1988).

For obtaining the high yielding hybrids, that show heterosis effects due to sets of genes responsible for yield and not for response, crossing among the divergent varieties within the cluster is recommended (Singh, 1988).

The following twelve genotypes were grouped in cluster I consistently in all the analyses. They are G 21, G 27, G 29, G 32, G 35, G 37, G 39, G 40, G 47, G 52, G 55 and G 58. Out of these twelve genotypes only five genotypes viz., G 27, G 32, G 35, G 39 and G 40 showed high mean seed yield. Added to this they also exhibited high mean performance for most of the characters that were positively and significantly associated with seed yield and seed size. By crossing among these genotypes high yielding genotypes with a response similar to the parents can be obtained.

Character contribution towards genetic diversity

Relative ranking of the contribution of sixteen characters towards genetic diversity based on D^2statistic, revealed high Genotype X Season interaction in their percentage contribution towards genetic diversity. Volume of 1000 seeds followed by oil content exhibited highest contribution towards genetic diversity in SI, SIII and pooled analysis. In SII, oil content followed by number of capsules exhibited the highest contribution. The highest contributor-volume of 1000 seeds scored eleventh rank in this season (SII) (Table 28). Eventhough oil content scored second rank in three seasons, its contribution was constant across seasons and pooled over seasons (Table 29).

Table 28. Percentage contribution of various characters to genetic divergence (Based on D^2 – Statistic)

S. No.	Characters	SI	SII	SII	P
1.	Days to maturity	05.54	07.85	07.97	09.21
2.	Height of the plant maturity	00.11	01.07	00.17	00.00
3.	Number of branches	02.82	00.40	00.00	00.51
4.	Number of flowers	03.90	09.15	05.76	09.04
5.	Number of capsules	02.43	18.42	08.47	04.18
6.	Volume of branches	03.50	12.09	02.15	06.21
7.	Number of seeds per capsule	00.17	01.69	00.28	00.11
8.	Length of seeds	07.68	12.43	01.02	04.92
9.	Breadth of seeds	02.09	04.12	00.00	00.00
10.	Weight of 1000 seeds	05.37	03.79	00.00	01.19
11.	Volume of 1000 seeds	47.12	01.64	32.66	41.19
12.	Density of seeds	07.18	00.00	00.90	05.31
13.	Oil content	09.83	22.20	26.27	15.71
14.	Seed yield	00.17	00.00	05.88	01.30
15.	Total dry matter production	01.24	04.07	04.41	01.02
16.	Harvest index	00.85	01.07	04.01	00.11

Table 29. Relative ranking of characters contributing towards diversity (Based on D^2 statistic)

Rank	SI	SII	SIII	Pooled
1	Volume of 1000 seeds	Oil content	Volume of 1000 seeds	Volume of 1000 seeds
2	Oil content	Number of capsules	Oil content	Oil content
3	Length of seeds	Length of seeds	Number of capsules	Days to maturity
4	Density of seeds	Volume of capsules	Days to maturity	Number of flowers
5	Days to maturity	Number of flowers	Seed yield	Volume of capsules
6	Weight of 1000 seeds	Days to maturity	Number of flowers	Length of seeds
7	Number of flowers	Breadth of seeds	Total dry matter production	Density of seeds
8	Volume of capsules	Total dry matter production	Harvest index	Number of capsules
9	Number of branches	Weight of 1000 seeds	Volume of capsules	Seed yield
10	Number of capsules	Number of seeds per capsule	Length of seeds	Weight of 1000 seeds
11	Breadth of seeds	Volume of 1000 seeds	Density of seeds	Total dry matter production
12	Total dry matter production	Height of the plant at maturity harvest index	Number of seeds per capsule	Number of branches
13	Harvest index	Number of branches	Height of the plant at maturity; weight of 1000 seeds	Number of seeds per capsule; harvest index;
14	Number of seeds per capsule; seed yield	Density of seeds; seed yield	Weight of 1000 seeds	Height of the plant at maturity; breadth of seeds
15	Height of the plant at maturity		Breadth of seeds; Number of branches	

4

Genetic Diversity in Blackgram [*Vigna mungo* (L.) Walp.]

Thirugnanakumar, S. and S. Thirumalraj

Introduction

India is the largest producer of pulses in the world, both in quantity and variety. But, presently, India is one of the largest importer of pulses. India imported pulse grains upto a tune of approximately 4000 crore rupees during 2006-07. It accounts for about 20 per cent of the total import bill during 2006-07. This is because of the green revolution (quantum jump in cereal production) accompanied with changes in the infrastructures and incentives including input supplies and price support systems in favour of major cereals. It altered the traditional cropping pattern against pulses with decades of emphasis on research as well as production in superior cereals coupled with a near total neglect of the rainfed areas, pulses were driven out of not only the irrigated areas, but also the rainfed farming and were relegated to the marginal lands. Further more, the weak infrastructural support for input supply, credit and marketing in the rainfed areas, the traditional home of pulses, also adversely affected the prospects for pulses. A vicious circle of low input use and low output thus got built in. Lack of processing and marketing facilities in pulses also contributed its share to the woes of pulse growers. The increasing shortage of pulses, resulted in the need for more and more import which, in turn has dampened the speed of rejuvenation of pulses sub-sector of agriculture sector.

The net per capita availability of pulses has fallen from 60.7 (1950-51) to 33.4 g (1991-92). On the contrary, the daily *per capita* availability of cereals has registered an increase from 334.2 (1950-51) to 443 g (1991-92). It leads to increase in the pulses/cereal ratio from 0.18 to 0.08 g *i.e.,* by 55 per cent. This is despite the increases of 24.36 per cent in area; 61.71 per cent in production and 29.93 per cent in production. It is significant to note that almost all these increases had taken place during 1950-51 to 1960-61. The area, production and yield of pulses have either stagnated or declined since then. This stagnation of pulse production and productivity should be viewed seriously not only from the point of security and quality of food for our people, especially the poor and the rural population, but also from the soaring import bill, imbalance and distribution in the cropping pattern and resource allocation among various sub-sectors of agriculture.

It is palpable that growth in production and productivity of pulses has lagged far behind the population growth during the past three decades. In addition, though the consumption expenditure on pulses is higher in the rural than in the urban areas, it has been falling over times in terms of percentage of both total expenditure and food expenditure, indicating a greater imbalance in the food habits of the rural population.

The requirement for pulses during this decade is more than 20 million tonnes. But, the production capacity will remain behind the projected demand (13.14 million tonnes, during 2006-07). Creating wide gap and causing deficiency in protein intake through pulses. Shortage of pulses will thus continue to be a cause of concern in the near future. The country has to make a much more vigorous efforts to bridge this gap and save the foreign exchange required to meet the demand through imports.

The genus *Vigna* (Family: Leguminosae) contains several species that are important in world agriculture. Urd beans, referred as blackgram [*Vigna mungo* (L.) Hepper], mung beans [*V. radiata* (L.) Wilczek] referred as greengram and cowpea [*V. unguiculata* (L.) Walp.] are grown on more than million hectares annually. They provide a significant portion of the dietary protein for many people, especially vegetarians. Additionally, several other *Vigna* species, *e.g.*, adzuki beans [*V. angularis* (Willd.) Ohwi and Ohashi], moth beans [*V. aconitifolia* (Jacq.) Marechal] and rice beans [*V. umbellata* (Thumb.) Ohwi and Chashi], are important in the diets of many of the societies.

The cultivated *Vigna* species share desirable features. They can be grown successfully over a wide range of environmental conditions and all provide inexpensive protein that can be prepared easily in a number of edible forms, such as tender green shoot and leaves, immature pods and green as well as dry seeds. The cultivated *Vigna* species also are valued as fodder, cover and green manure crops.

Urd bean is grown throughout South East-Asia and is a staple grain legume crop in India. It originated from India. Urd bean is the most widely cultivated pulse crops in the country. It is grown over on an area of 30,11,300 hectares with a production of 12,95,400 tonnes during 2000-01 in India. The average productivity of urd beans is 431 kg per hectare. But, the target is 702 kg per hectare. About 18 per cent of the total blackgram area is in Andhra Pradesh.

Andhra Pradesh accounts 30 per cent of the total blackgram production in India. In Tamil Nadu, blackgram was cultivated in an area of 276 thousand hectares (9 per cent, during 2000-01) and production was about 127 thousand tonnes (10 per cent) during 2000-01. In Tamil Nadu, the productivity of urd bean is 423 kg per hectare, during 2000-01. The target is to achieve 502 kg per hectare. Concerted efforts from all the agricultural scientists is needed to achieve the goal.

Blackgram [*Vigna mungo* (L.) Hepper] is an important pulse crop cultivated during monsoon, post-monsoon and summer season as a sole crop as well as inter-crop in India. It is self-pollinated, diploid and cleistogamous upto 42 per cent (Bose, 1932; Narasimman, 1929; Daniel Sundararaj and Thulasidas, 1976). Emasculation and hybridization techniques have been described by Sen and Jana (1963). The chromosome number of blackgram is 2n=22 (Chaurasia and Sharma, 1974; De and Krishna, 1966).

Blackgram and greengram, earlier included in the genus *Phaseolus,* now belong to the genus *Vigna* of the tribe *Phaseolae*. Based on morphological characters, such as nature of stipules, curvature of keel petals, nature of pods and biochemical evidences, it was revised by Verdicourt (1970) in the *Phaseolus* Linn. and *Vigna* Savi. The yellow flowered species of *Phaseolus* under the subgenus Ceratotropics piper were transferred to the genus *Vigna* subgenus *verdi*. The genus *Vigna* is represented by 100-150 species mainly found in Africa and India (Purseglove, 1968).

Simple selection techniques were successful in breeding crop plants for a long time. But many generations of deliberate selection has resulted in materials that are no more amenable for further advance (Williams, 1964). This necessitates the need for a refined method for choice of parents (Hayward, 1979). Blackgram being a self-pollinated crop, traditionally additive variation has been exploited in breeding high yielding purelines.

A number of genetic studies, however, have shown the presence of significant non-fixable genetic variation in this crop. The growth and development of this crop are highly influenced by environment (Singh *et al.,* 1976; Sood and Saini, 1989). It is commonly observed that the relative performance of different genotypes varies in different environments.

Blackgram is used in preparation of fermented foods. Batra and Milner (1976) have reviewed the Asian fermented foods and beverages prepared from pulses.

Thirumal Raj (2008) studied 25 genotypes of blackgram and measured genetic diversity. The mean values of the 25 genotypes were transformed into standardized uncorrelated mean values. The D^2 values were computed for all possible pairs $\frac{25(25-1)}{2} = 300$ of genotypes, and are presented in Table 31.

The highest D^2 value was observed between the genotypes G_8 and G_9 (113.428) and the lowest values was registered between the genotypes G_4 and G_{10} (1.683).

By the application of clustering technique, the 25 genotypes were grouped into six clusters (Table 32). Cluster I and cluster IV each consisted of four genotypes. The cluster II and cluster V each consisted of two genotypes. Cluster III consisted of three genotypes. Cluster VI consisted of ten genotypes.

D^2 analysis of the 25 genotypes confirmed the presence of high diversity among the genotypes by their resolution into as many as six clusters. The genotypes of different eco-geographical origin were grouped in a single cluster as well as in different clusters.

The cluster I consisted of four genotypes each one originating from Erode Local, Sankari Local, Kandiamedu and Uttar Pradesh, respectively. Clusters II consisted of two genotypes, each one originating from Chidambaram Local, Andhra Pradesh, respectively. Cluster III consisted of three genotypes each one hailing from Chidambaram Local, Aduthurai and Sirkali. Cluster IV consisted of four genotypes, two originating from Chidambaram and one each originating from Madurai and Salem. Cluster V consisted of two genotypes each one originating from Salem and Paiyur, respectively. Cluster VI consisted of ten genotypes, three originating from Vamban, two originating from Coimbatore and each one originating from Chidambaram, Sethiyathoppu, Hathipatti, Maharastra, Tindivanam and Paiyur, respectively. The result indicated the presence of wide range of diversity among the types of the present study. It is in confirmity with the findings of Sharma and Misra, (1997), Bisht *et al.* (1999), Verma *et al.* (1997). This genotypes could well be exploited in recombination breeding.

The tendency of genotypes from diverse geographic region to group together in one and the same cluster or scattered distribution of genotypes of same geographical origin in to group different clusters had been observed in the present study. The studies of several worker confirmed the tendency, that the geographical distribution and genetic diversity would not be directly related in any of the crops, tobacco (Murthy, 1965), linseed, wheat, sorghum (Murthy and Arunachalam, 1969), cotton (Singh and Gupta, 1965), safflower (Rao *et al.,* 1980) and sesame (Thirugnanakumar, 1991).

Genetic drift and selection in the different environment could cause greater diversity than geographical distance (Murthy, 1965). However, geographical isolation could be important in the alteration of breeding structure as pointed out by Wallace (1963).

Cluster II including two genotypes showed the least intra cluster divergence. This indicated the closeness of the genotypes G_7 and G_{16}. This may be explained on the basis that yield being a complex character of polygenetic inheritance, similar phenotypes could be produced by many different combination of genes and such combination may have similar selective advantage (Singh and Gupta, 1968). Further, under constant selection on the segregating populations such similar types are excepted to be a established.

The intra cluster distance was maximum in the cluster IV. It included four genotypes namely, G_5, G_6, G_{15} and G_{17}. This may indicate that these genotypes differ genetically. The limited gene exchange between the genotypes or selection for diverse characters could be responsible for such high intra cluster divergence.

The inter cluster distance was maximum between II and III clusters followed by I and III clusters. The cluster II consisted of four genotypes and cluster III also consisted of three genotypes. The cluster II consisted of two genotypes. Cluster II recorded higher mean for number of clusters per plant. Cluster III recorded higher number of branches per plant, number of pods per cluster, number of pods per plant, number of seeds per pod, length of pod, 100 seed weight and seed yield per plant. Hence, crossing among the genotypes which were grouped in these clusters would be useful in evolving high yielding recombinant genotypes (Table 33).

As pointed out by Sokal (1965), the choice of character is important in multivariate analysis and their choice made in the present study appears to be appropriate. Cyclic breeding and crossing may be helpful in bringing new genes in to a population of blackgram and thus, expanded the range of adaptation. As brought out by Mather (1943) polygenic variability which is necessary for prospective, adaptative and evaluating change need not exist as true phenotypic variation which effect fitness. Therefore, phenotypic uniformity with in the population appears to be very useful in blackgram. In fact, this is possible with the proper use of the suggested parents in the formation of gene complexes to replace the existing verities.

It can be concluded that there is no relationship between genetic and geographical diversity in the material under investigation. Inter and Intra clusters distances were also not related.

Table 30. Salient features of the blackgram genotypes studied

S. No.	Genotypes	Seed colour	Origin
G_1	Erode Local	Black colour	Erode
G_2	Sankari Local	Black colour	Salem
G_3	Chidambaram Local	Black colour	Chidambaram
G_4	Kandiamedu	Black colour	Chidambaram
G_5	Mandapam	Black colour	Chidambaram
G_6	Poolamedu	Black colour	Chidambaram
G_7	Nanjaloor Local	Black colour	Chidambaram
G_8	ADT 3	Black colour	Aduthurai
G_9	Vanniyur Local	Black colour	Chidambaram
G_{10}	T 9	Black mottled	Uttar Pradesh
G_{11}	Sirkali Local	Black colour	Sirkali
G_{12}	Sethiyathoppu	Black colour	Sethiyathoppu
G_{13}	Hathipatti Local	Black colour	Hathipatti
G_{14}	VBN 1	Black colour	NPRC Vamban
G_{15}	Madurai Local	Black colour	Madurai
G_{16}	Ongole Local	Black colour	Andhra Pradesh
G_{17}	Omalur Local	Black colour	Salem
G_{18}	Maharastra Local	Black colour	Maharastra
G_{19}	Mettur Local	Black colour	Salem
G_{20}	Paiyur 1	Black colour	Paiyur
G_{21}	CO 1	Black colour	Coimbatore
G_{22}	CO 2	Black colour	Coimbatore
G_{23}	VBN 2	Black colour	NPRC Vamban
G_{24}	TMV 1	Black colour	Tindivanam
G_{25}	VBN Culture RTU 16	Black colour	NPRC Vamban

Table 31. D^2-values

	1	2	3	4	5	6	7
1	0.000	3.456	110.020	12.695	6.035	67.315	7.956
2		0.000	106.350	11.954	7.531	60.530	5.706
3			0.000	55.405	82.473	21.81	77.421
4				0.000	6.029	24.446	4.160
5					0.000	46.703	3.704
6						0.000	39.196
7							0.000

	8	9	10	11	12	13	14
1	109.543	10.480	12.529	97.521	18.334	32.775	44.173
2	107.543	13.121	10.084	94.754	21.837	31.189	44.474
3	5.218	106.236	55.679	4.826	48.600	30.642	36.352
4	60.092	14.167	1.683	51.304	4.426	5.536	13.594
5	83.768	12.343	5.047	75.941	11.493	19.727	24.984
6	23.348	68.920	25.851	20.107	28.677	10.358	21.400
7	82.199	13.195	2.866	69.461	9.079	16.540	20.767
8	0.000	113.428	57.686	3.261	53.466	35.447	48.706
9		0.000	15.676	99.855	16.469	28.165	35.149
10			0.000	48.918	5.169	7.894	16.414
11				0.000	43.032	28.911	38.043
12					0.000	7.357	10.988
13						0.000	8.219
14							0.000

	15	16	17	18	19	20	21
1	35.308	6.988	25.223	64.133	72.685	73.195	13.417
2	30.912	8.700	24.423	60.153	72.442	69.699	12.751
3	48.988	83.076	50.981	16.957	18.758	17.369	52.839
4	12.670	5.799	7.926	23.916	30.876	30.506	4.924
5	18.937	2.689	13.646	44.813	49.256	53.570	8.191
6	24.846	48.545	21.383	13.320	18.106	16.656	35.687
7	11.199	1.939	8.117	38.079	41.973	42.077	7.727
8	59.108	89.302	54.743	25.000	26.088	27.217	57.682
9	38.725	9.755	34.234	51.628	62.832	65.531	16.070
10	11.254	5.473	6.905	25.350	29.702	29.995	4.785
11	46.352	77.190	44.420	16.507	19.101	17.320	47.705
12	13.200	7.772	7.937	21.450	20.532	20.812	7.618
13	15.434	19.834	12.332	7.978	15.958	16.859	11.036
14	8.375	20.894	12.623	10.799	13.163	16.290	17.893

15	0.000	14.338	3.814	24.713	21.450	20.306	16.914
16		0.000	10.748	43.355	43.445	45.734	8.904
17			0.000	28.300	22.295	21.716	15.166
18				0.000	11.622	12.929	24.379
19					0.000	5.495	36.448
20						0.000	35.562
21							0.000

	22	23	24	25
1	11.127	43.527	26.808	37.800
2	14.206	39.882	24.939	38.987
3	102.098	31.411	35.446	31.356
4	13.665	12.491	6.230	14.413
5	11.887	29.163	14.773	22.099
6	63.544	12.806	20.105	20.205
7	6.826	18.530	10.127	18.417
8	111.403	41.083	42.416	32.803
9	10.154	41.360	28.718	39.953
10	13.758	13.118	6.062	11.141
11	93.481	29.758	31.461	24.811
12	11.720	10.172	5.767	7.690
13	30.046	7.621	6.344	10.727
14	28.025	8.542	6.482	11.042
15	21.037	7.437	5.232	10.029
16	3.796	22.484	12.787	19.736
17	17.385	7.309	6.030	7.111
18	53.991	13.395	13.599	16.669
19	51.102	9.125	18.860	8.223
20	51.451	7.547	16.705	12.908
21	18.217	19.796	5.653	16.503
22	0.000	27.635	21.167	27.821
23		0.000	7.173	7.513
24			0.000	6.867
25				0.000

Table 32. Composition of D^2 clusters

Cluster	Number of genotypes	Name of genotypes
I	4	G_1, G_2, G_4, G_{10}
II	2	G_7, G_{16}
III	3	G_3, G_8, G_{11}
IV	4	G_5, G_6, G_{15}, G_{17}
V	2	G_{19}, G_{20}
VI	10	G_9, G_{12},G_{13}, G_{14}, G_{18}, G_{21}. G_{22}, G_{23}, G_{24}, G_{25}

Inter and intra-cluster distance

The inter and intra-cluster D^2 and D-values among the ten clusters are presented in Table 33. Cluster II had the lowest intra-cluster generalized distance of (1.392) and maximum intra-cluster distance was recorded by cluster IV (4.643). The inter-cluster distance ranged from 5.956 (between cluster I and cluster II) to 79.775 (cluster II and cluster III). The inter cluster distance between clusters I and II was also high (79.568) (Fig. 1).

Cluster mean values of the characters

The cluster means for all the nine characters in respect of six clusters were estimated and are given in Table 34.

Cluster V exhibited minimum value for plant height (27.51 cm). Cluster III recorded maximum mean number of branches (2.77), cluster II registered higher mean number of cluster per plant. Cluster III evinced higher mean for number pods per cluster (5.41), number of pods per plant (34.21), number of seeds per pod (5.44), length of pod (4.96), 100 seed weight (4.90) and seed yield per plant (8.08).

Contribution of characters towards diversity

Seed yield per plant contributed maximum towards genetic diversity (55.00 per cent) followed by 100 seed weight (31.00) (Table 35).

Table 33. Inter and intra-cluster distance extent diversity among the cluster in blackgram

Clusters	I	II	III	IV	V	VI
I	8.733 (2.955)	5.956 (2.44)	79.568 (8.920)	22.338 (4.726)	51.137 (7.151)	20.921 (4.574)
II		1.939 (1.392)	79.775 (8.932)	17.317 (4.161)	43.307 (6.581)	16.430 (4.053)
III			4.435 (2.106)	51.003 (7.147)	20.975 (4.580)	50.231 (7.087)
IV				21.55 (4.643)	27.919 (5.280)	20.100 (4.483)
V					5.495 (2.344)	25.223 (5.022)
VI						17.679 (4.205)

Table 34. Cluster means of 25 blackgram genotypes

Clusters	X_1	X_2	X_3	X_4	X_5	X_6	X_7	X_8	X_9
I	29.30	2.58	7.20	4.22	30.08	4.50	4.60	4.40	6.35
II	28.03	2.50	7.38	4.28	31.11	4.67	4.69	4.39	6.88
III	30.77	2.77	6.42	5.41	34.21	5.44	4.96	4.90	8.08
IV	27.90	2.50	7.28	4.48	31.63	4.50	4.56	4.58	6.73
V	27.51	2.33	5.70	5.10	29.48	4.83	4.84	4.76	6.72
VI	28.34	2.67	6.87	4.46	31.47	4.70	4.75	4.55	6.69

Table 35. Percentage contribution of various characters to genetic divergence (based on D^2 statistics)

Characters	Number of first rank	Percentage contribution
Plant height at maturity	11	3.66678
Number of branches	3	1.0000
Number of clusters per plant	3	1.0000
Number of pods per cluster	2	0.6667
Number of pods per plant	13	4.3333
Number of seeds per pod	2	0.6667
Length of pod	6	2.0000
100 seed weight	93	31.0000
Seed yield	167	55.0000

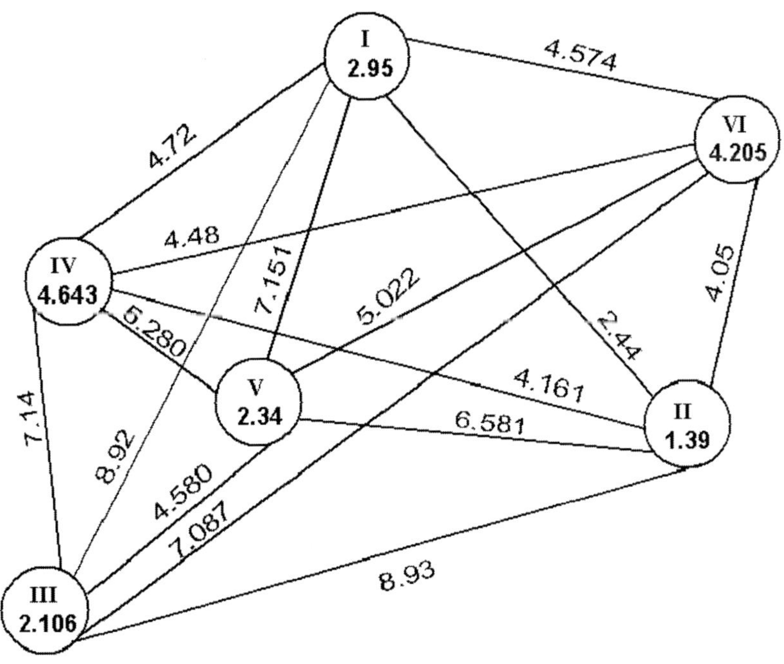

Fig. 1. Cluster diagram (*Not to scale)

5

Genetic Diversity in Greengram [*Vigna radiata* (L.) Walp.]

Thangavelu, P. and Manamalli

Introduction

The nature and magnitude of genetic variability present in the genetic stock afford the plant breeder, information to formulate a purposeful breeding programme for crop improvement. The association and the influence of yield components with yield, their heritability and genetic advance would characterise the various steps in the breeding programme in achieving the aim of the breeder. Studies on these lines in various crops have hastened the realization of crop improvement in general.

India grows a variety of pulse crop which probably no other country in the world grows. These are multiple utility, energy high crops as they provide protein for human consumption, green nutritious fodder and enrich soil fertility through biological nitrogen fixation. *Vigna radiata* (L.) Wilczek, commonly known as greengram or mungbean, is the most widely distributed of the six cultivated Asiatic Vigna species.

The current breeding approaches to yield improvement in greengram are based on the recognition that this crop have lost a large part of their genetic variability in the process of adaptation to stress environment. This variability must now be regenerated.

The foremost need for this purpose is the collection of germplasm from different parts of the world and the organisation of a very intensive hybridization programme, using diverse genetic stocks as parental lines. Selection of parents for hybridization based merely on geographic diversity can no more be a suitable criterion. The importance of genetic diversity has long been appreciated by crop specialists. Such an analysis will eventually help to choose desirable parents for hybridization and evolve superior genotypes.

Also, selection of parents based on yield alone is not likely to be efficient. Indeed a knowledge of genotypic correlation of yield components with each other and the yield, and the relative contribution of the component characters towards yield, indicate those components upon which greater emphasis should be placed by the breeder for the purpose of selection.

Lack of a concept of morpho-physiological framework has lead to high seed yield rather than genetic variability. Hence, it is important to gather information on the physiological basis of yield variation and this can help in outlining a suitable morpho-physiological framework related to higher seed yield.

Understanding the magnitude of genetic divergence for yield components is very important in planning breeding programme of crop plants. It helps both in choosing parental combinations to get maximum recombinations and also in understanding the pattern of evolution for different characters. Singh and Gupta (1968), using D^2 statistic in cotton found that the progenies derived from diverse crosses gave divergent and useful progenies. Since then works have been carried for analyzing the genetic divergence and planning the breeding programme based on the extent of divergence. This chapter focused on the quantification of the magnitude of genetic divergence for yield and its morphological characters in green gram.

Manamalli (1997) studied genetic diversity in 28 genotypes of greengram. Twenty eight genotypes fell into ten clusters. An examination of the distribution of the genotypes over the clusters (Table 36) revealed that there was no clear relationship between clustering pattern of the genotypes and their geographical sources. In many cases, the genotypes from different geographic sources fell in the same clusters. This could be appreciated in the clusters I, IV, V, VI and VIII. Similarly, the genotypes from similar geographic sources had a tendency to spread over different clusters. For example, the three Vamban genotypes VGG 4, VGG 7 and VGG 8 spread over the clusters I, V and VI respectively.

The results suggested that geographic isolation is not the factor causing genetic diversity and this point should be considered in selecting parents for hybridization programme. To explain the lack of relationship between geographic and genetic diversity, many interpretations were made by different authors.

Murthy and Arunachalam (1966) and Upadhya and Murthy (1970) pointed out that genetic drift and natural selection in different environments could cause high diversity among the races than geographic isolation can do. Das and Borthakur (1973) and Dhagat and Singh (1983), however, viewed that if similar selection criteria are exercised in different geographic areas, it may result in genetic similarity of the cultivars of different geographic regions. All these factors can be considered to support the current finding. These factors lay stress on the fact that in breeding programmes, geographic diversity need not be considered always as an index of diversity for the selection of parents.

Cluster VI had the maximum intracluster distance (Table 37) suggesting that considerable divergence existed within a cluster. Maximum intercluster distance was recorded between the clusters IV and X. Also, cluster IV and VI, IV and VII, I and IV and IV and III and IV were considered to be highly divergent groups. The types in these clusters may serve as potential parents and crossing among them may result in the heterotic expression for yield components. Wilcox and Wilsie (1964), Ramanujam *et al.* (1974) and Natarajan *et al.* (1988) also reported that heterosis can be exploited by the genetic diversity of the germplasm.

Cluster II had increased mean values for number of clusters per plant and number of pods per cluster. Cluster III recorded high mean values for number of branches. Cluster X registered high mean value for plant height and the cluster IV recorded high mean values for number of seeds per pod, hundred seed weight and seed yield per plant (Table 38). Intercrossing the types from these clusters might result in wide array of variability for exercising effective selection.

An assessment of the contributions of the different characters (Table 39) revealed that hundred seed weight contributed maximum to total divergence followed by days to flower, number of pods per cluster, number of seeds per pod, number of branches per plant, number of clusters per plant, seed yield per plant and plant height in the order. Of these eight characters, the first four contributed about 77 per cent to the total divergence, indicating their major role in maintaining variability in the natural population. The same was recorded by Natarajan *et al.* (1988). The genetic divergence offered by these characters will offer scope for improving yield through rational selection.

Variations in leaf shape viz., lobed leaflet may provide as a marker character in the selection of genotypes and variations in seed colour and seed size help to identify the cultivar. Bold seeded type could contribute to increase the yield if incorporated into a good yielding genotype but possessing small sized seeds.

Table 36. Composition of D^2 clusters in greengram

Cluster Numb	Number of genotypes	Name of the genotypes	Source/Origin
I	13	EC	Taiwan
		WGG 48	Warangal
		GPLM 139	Vamban
		PDM 89.226	Punjab
		PUSA 9032	New Delhi
		AVRDC 5486	Taiwan
		ML 267	Ludhiana
		PLM 214	New Delhi
		GPLM 163	Vamban
		EC 309459	Taiwan
		VGG 4	Vamban
		PLM 404	New Delhi
		EC 314292	Taiwan
II	1	WGG 42	Warangal
III	1	WGG47	Warangal
IV	3	EC 310277	Taiwan
		EC 312282	Taiwan
		LM 608	Ludhiana
V	2	VGG 7	Vamban
		KM 2	Kudumianmalai
VI	3	VGG 8	Vamban
		PLM 696	New Delhi
		ADT 3	Aduthurai
VII	1	GPLM 83	Vamban
VIII	2	PLS 264	Punjab
		EC 314295	Taiwan
IX	1	PDM 89.227	Punjab
X	1	CO 4	Coimbatore

Table 37. Average intra (diagonal) and intercluter D^2 and D (bold values) in greengram

Cluster	I	II	III	IV	V	VI	VII	VIII	IX	X
I	69.78 8.35	382.57 19.56	291.44 17.07	1927.11 43.90	234.37 15.31	153.96 12.41	163.25 12.78	654.32 25.88	179.27 13.39	292.50 17.10
II		0	22.77 5.27	1745.71 41.78	256.36 16.01	295.14 17.18	170.55 13.06	828.35 28.78	218.81 14.79	454.60 21.32
III			0	1611.52 40.14	159.35 12.62	241.74 15.55	171.1 13.08	677.14 26.02	217.13 14.74	446.24 21.12
IV				37.38 9.35	1102.16 33.20	2420.11 49.19	2261.43 47.55	455.30 21.34	1654.23 40.67	2883.65 57.70
V					48.29 6.95	344.75 18.57	323.94 17.99	271.48 16.48	237.77 15.42	699.13 26.44
VI						91.67 9.57	137.18 11.71	991.76 31.49	208.17 14.43	294.31 17.16
VII							0	952.91 30.87	160.36 12.66	126.44 11.24
VIII								86.85 9.32	596.81 24.43	138.10 37.26
IX									0	361.62 19.02
X										0

Table 38. Cluster mean for yield attributing morphological characters in greengram

Cluster	Days to first flower	Plant height	Number of branches per plant	Number of clusters per plant	Number of pods per cluster	Number of seeds per pod	Hundred seed weight (g)	Seed yield per plant (g)
I	40.21	24.84	3.41	4.35	2.71	9.58	3.50	3.97
II	37.27	25.62	3.33	8.13	5.06	10.13	3.50	14.59
III	36.27	27.91	3.93	7.80	4.45	10.60	3.53	13.13
IV	40.09	31.79	3.33	5.40	3.52	10.71	7.39	15.13
V	36.40	27.40	3.77	6.03	3.29	9.50	4.32	8.54
VI	36.13	23.69	3.27	4.98	3.63	10.18	2.92	5.19
VII	41.33	27.29	3.27	6.93	3.98	9.33	3.27	8.46
VIII	39.13	28.18	3.80	4.37	2.91	10.07	5.92	7.75
IX	40.13	23.34	3.27	3.87	4.77	9.67	4.10	7.42
X	45.73	32.38	3.73	6.27	3.70	10.60	2.83	4.79

Table 39. Rank total and the relative contribution of yield attributing morphological characters towards genetic divergence

S. No.	Character	Rank total	Contribution (%) of each character
1	Days to first flower	69	18.25
2	Plant height	11	2.91
3	Number of branches per plant	37	9.80
4	Number of clusters per plant	25	6.61
5	Number of pods per cluster	69	18.25
6	Number of seeds per pod	42	11.11
7	Hundred seed weight	113	29.89
8	Seed yield per plant	12	3.18
		378	

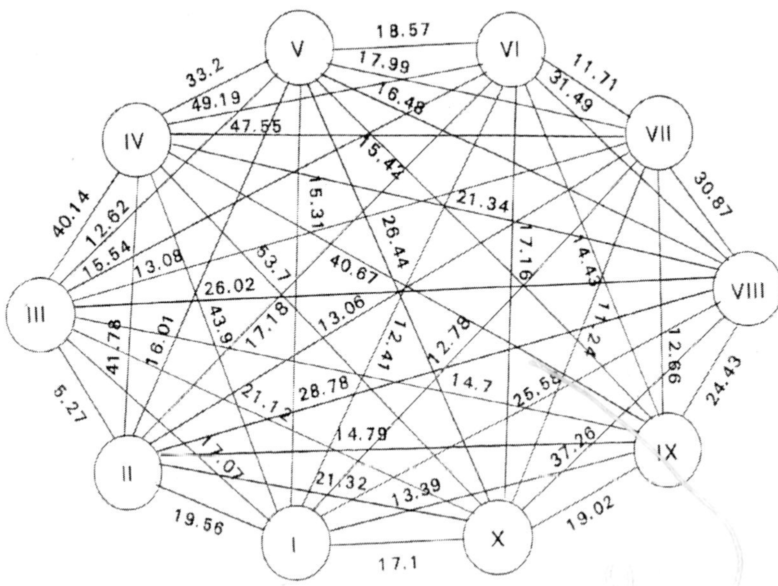

Fig. 2. Clustering pattern based on morphological characters (not to scale) in greengram

6

Genetic Diversity in Bhendi [*Abelmoschus esculentus* (L.) Moench]

Senthilkumar, N. and K. Ramya

In recent years, we are just experiencing a surplus production in cereals and oil seeds leading to self-sufficiency through green revolution and yellow revolution. However, shortage in the production of vegetables has drawn the attention for increasing the cultivation of vegetables to provide food and nutrition security.

Vegetables are being cultivated over an area of 49 million hectares with the production potential of 487 million tonnes in the world. This accounts for 2.85 per cent of total cropped area in the world. India is the second largest producer of vegetables next to China, sharing nearly 12 per cent of the total world output (Joshi and Shukla, 1997). Vegetables occupy nearly 8.02 million hectares with the production potential of 133.10 million tonnes in the year 2009 in India (Anonymous, 2009).

Bhendi (*Abelmoschus esculentus* (L.) Moench) is an important member of family malvaceae and commonly known as okra or ladyfinger in India. It is one of the most ancient and traditional vegetable crops grown in tropical and subtropical regions of the world (Martin and Ruberte, 1978). Bhendi is originated from tropical Africa with 2n = 8x = 72 or 130 or 144 chromosomes and is an allopolyploidy in nature (Joshi and Hardas, 1956; Suresh Babu, 1987). Martin (1982) reported that it behaves as a diploid. Under the genus *Abelmoschus,* there are 30 species in the old world and four in the new world (Joshi *et al.,* 1974). Out of them, *Abelmoschus esculentus,* is the only species known to be cultivated. Bhendi is being raised in India in an area of 0.44 million hectares with an average production of 4.52 mt with the productivity of 10.27 t/ha (Anonymous, 2009).

The production of green tender fruit was 35.24 lakh tonnes from major producing states in India namely Uttar Pradesh, Bihar, Orissa, West Bengal, Andhra Pradesh, Karnataka and Assam. In Tamilnadu, bhendi is cultivated in an area of 5,160 ha with an annual production of 38,227 metric tonnes (FAO, 2007). Further, bhendi has a vast potential on one of the foreign exchange earner and accounts for about 60 per cent of the export of fresh vegetables excluding potato, onion, garlic etc. The importing countries are Middle East, Western Europe and USA (Anonymous, 1988). Green tender fresh fruit when stored at 7 – 10 °C of temperature, at 90 – 95 per cent of relative humidity remain fresh for 7 – 10 days.

Being native of tropical Africa, it is a prized vegetable of India. The young, immature, non-fibrous, edible and tender delicious fruits of bhendi used as vegetables in salads, soups and sauces, are also good sources of iodine, calcium, iron and vitamin A, B and C (Woodroof, 1927; Ephenhuijsen, 1974, Purseglove, 1974; Cobley and Steele, 1976; Martin and Ruberte, 1978). Bhendi leaves are most frequently cooked as spinach or added to soups and stews. It provides vitamins A and C, protein, calcium and iron (Irvine, 1952). It has been reported that bhendi has an average nutritive value (ANV) of 3.21, which is higher than tomato, egg plant and most of the cucurbits except bittergourd (Grubben, 1977). Matured and dried seeds of bhendi are roasted and ground as a coffee substitute, or are added to coffee as an adulterant (Burkill, 1935). Bhendi dry seeds contain 13 to 22 per cent oil. The percentage of oleic acid and linoleic acid is 60 to 65 per cent. Residue of the whole seed contains 18 to 27 per cent protein and relatively high appeared to be a suitable ingredient in diet for rats and chicks (Edwards and Miller, 1947). The dried bhendi is a processed product for preservation and export. Bhendi seed is rich in tryptophan (94 mg/g N) and has an adequate content of sulphur containing amino acid (189 mg/g N). The protein of bhendi could thus complement that of legumes or cereal grain in some dietary combinations (Savello *et al.,* 1980).

The bhendi plant is used for clarification of sugarcane juice before it is converted into jaggery. The medicinal value of bhendi is associated with genitourinary disorders, sprematorhoea and chronic dysentery. Bhendi, among other plants may serve as gossypol source or bhendi seed in the diet may reduce fertility. Gossypol is being tested as an orally administered contraceptive in China. The young fruits also have high mucilage content and a mucilaginous preparation from then named 'bhendi-mucilage F' by Tomoda *et al.* (1980), could be used as a plasma replacement or blood volume expander (Benjamin *et al.,* 1951 a and b; Purseglove, 1974). The mucilage is potentially useful as an extender of serum albumin and as an egg white extender or substitute (Woolfe *et al.,* 1977). Bhendi is also a best fibre plant, but the fibre is inferior to those of corchorus and is therefore not extracted commercially. They are nevertheless important for many local uses (Purseglove, 1974; Cobley and Steele, 1976). The fibre of bhendi is a good material for making paper (Nelson *et al.,* 1961).

The per capita availability of vegetables is only 180 g against minimum requirement of 300 g for adults for a balanced diet. Within this view, we must increase the vegetable production in India to meet the requirements of our growing population (Dwivedi *et al.,* 2003). So it is necessary to increase the production and productivity to meet the self sufficiency of our daily requirements.

A further increase in bhendi productivity needs intensive research in genetics and crop breeding. Greater genetic diversity in the germplasm enhances the breeding potential and scope for crop improvement. Crosses between genetically diverse parents are likely to produce high heterotic effects and also produce a wide spectrum of variability in segregating generations. The Mahalanobis D^2 statistics is a very useful tool for estimating such genetic divergence.

The *per se* performance of the parents themselves does not always give a correct indication of their breeding potentialities. Breeders need to examine whether productivity is enhanced mainly by genes favoured by heterozygosity or homozygosity. Crop improvement through hybridization will be effective, if the information on the genetic architecture of the experimental population is available. The quantitative characters which are controlled by a large numbers of genes, which exerts very small effect, cannot be easily depicted through simple biometrical techniques. Indeed knowledge of combining ability and heterosis for fruit yield and its component characters should be placed greater emphasis, for improvement of this crop. Hence, potential diallel mating design proposed by Hayman (1954a and b) was adopted along with five parameter model of Hayman's (1958) generation mean analysis to find out the genetic architecture of bhendi in relation to fruit yield and its component traits.

Mahalanobis (1936) originally developed the concept of D^2 statistics. He used this technique in the study of anthropometry and psychometry. Rao (1952) suggested the application of this technique for the assessment of genetic diversity in the various breeding materials during plant breeding programme. Importance of genetic divergence in selection of parents for hybridization was stressed by many workers. According to Singh and Gupta (1968), the more diverse parents within a reasonable range, the more would be the chance of improving the trait. Usually in most of the conventional heterosis breeding programmes, geographic and phenotypic diversity are taken as the criteria for choosing genetically divergent populations and for isolation of inbred lines. Hybrids between lines of diverse origin, generally, display a greater heterosis than those between closely related parents. This has been observed in maize, alfalfa, cotton and several other crops. Generally geographic diversity has been considered as an index of genetic variability in crop plants. However, this may not be true for every case, as many workers postulated that geographic diversity need not necessarily be related to genetic diversity. Validity of the above assumption depends upon the association between geographic diversity and genetic diversity (Singh and Bain, 1968). Phenotypic divergence in a population has also been consisted as an index and criterion of genetic diversity (Rai, 1979).

Hazra *et al.* (2002) studied the genetic divergence of twenty two bhendi genotypes using D^2 analysis. They grouped the genotypes into five clusters. The first cluster contained sixteen genotypes and the second contained three, while third, fourth and fifth clusters contained one genotype each. They observed that grouping of genotypes in the same cluster confirmed that there was no parallelism between genetic diversity and geographical distribution.

Bendale *et al.* (2003a) grouped twenty five bhendi genotypes into eight clusters using Mahalanobis D^2 statistics to understand the nature of divergence and to assess the importance of a set of quantitative character related to economic yield in genetic differentiation. It was found that geographical diversity need not necessarily be related to genetic divergence.

Twenty two strains of bhendi were grouped into seven clusters. Cluster I, II, III and IV was the largest and together accounted for more than two third of total population. Genetic diversity was not related to geographical diversity (Dhaduk *et al.,* 2004).

Kiran Patro *et al.* (2005) evaluated forty one bhendi genotypes to estimate genetic divergence using Mahalanobis D^2 statistics. The forty one genotypes were grouped into eight clusters. Genotypes in cluster II had maximum intra-cluster distance. Inter-cluster distance indicated cluster I and VI followed by III and VI as the source of prospective parents for hybridization programme.

Mamta Kumari and Choudhury (2006) evaluated forty genotypes of bhendi in eleven clusters using seven characters by Mahalanobis D^2 statistics. Cluster I had maximum with 10 genotypes, while cluster II had eight genotypes and cluster III had seven genotypes. The maximum contribution towards genetic divergence was made by the trait fruit length (43.08 per cent).

Dhankhar *et al.* (2008) observed highly significant variability among ninety three genotypes. The genotypes were grouped into eight clusters. Cluster I had 24 genotypes, while cluster II had 15 genotypes and cluster III and cluster IV had 13 genotypes each. Intra-cluster distance was maximum in cluster V and the inter-cluster distance was maximum between cluster V and IV (8.26).

Jag Paul Sharma *et al.* (2008) studied the genetic diversity in thirty nine genotypes of bhendi for eleven characters. The genotypes were grouped in to eight clusters. Cluster I had 14 genotypes, while cluster II and III had eight and five genotypes, respectively. No relationship was found between clustering pattern and geographical distribution.

Singh and Singh (2008) studied genetic diversity among thirty four genotypes of bhendi using Mahalanobis D^2 statistics. The genotypes were grouped into seven clusters. Distribution of genotypes within various clusters were independent of geographical distribution.

Divergence analysis

Genetic divergence as measured by Mahalanobis (1936) in terms of generalized distance (D^2) has been one of the important statistical tools to estimate the genetic distance between any two genotypes. It provides rational basis for selection of parents for hybridization programme. Crosses among divergent parents are likely to yield desirable combinations. Therefore, a crossing programme can be initiated between the genotypes belonging to different clusters.

In this context, two important points are to be considered as (i) Choice of the particular cluster from which genotypes are to be used as parent in crossing programme and (ii) Selection of particular genotype from selected groups. The greater the distance between two clusters, wider is the genetic diversity among the parents, to be included in hybridization programme. Parents combining high yield potential with wider genetic diversity are likely to yield superior segregants within a short period.

In the present study, the 35 genotypes were grouped into ten different clusters, based on relative magnitude of D^2 values. Cluster II is the largest comprising of 12 genotypes, followed by cluster V with six genotypes and cluster VIII with four genotypes. The clusters namely II, III, IV, VI, VII and IX had two genotypes each and cluster X was monogenotypic cluster (Table 40).

The clustering pattern in the present study revealed that the genotypes from different sources clustered together showing that there was no association between ecogeographical distribution of genotypes and genetic divergence. Similar findings were also reported by many workers (Hazra *et al.*, 2002; Dhaduk *et al.*, 2004; Dhankhar *et al.*, 2008; Jag Paul Sharma *et al.*, 2008). This indicated that selections have been towards the same goal in the different centres of origin of these genotypes and yet, there is sufficient genetic variability that distinctly differentiates them into ten clusters. Hence, the chosen genotypes used in the present study could be considered as a valid material.

The intra-cluster distance varied from 0.00 to 29.96. Inter-cluster distance was minimum between cluster IV and IX suggesting closer relationship. Maximum inter-cluster distance existed between cluster VII and X followed by between cluster IV and X, between cluster III and VII, between cluster IX and X and between cluster III and IV. There existed sufficient inter-cluster distances between all the ten clusters. This clearly indicated that the genotypes included in these clusters are having broad spectrum of genetic diversity (Table 41).

The superior genotypes from the ten different clusters may be used as parents in hybridization programme to get superior hybrid combination as well as desirable segregants (Hazra *et al.,* 2002; Bendale *et al.,* 2003a; Kiran Patro *et al.,* 2005a; Singh and Singh, 2008; Jag Paul Sharma *et al.,* 2008).

Perusal of data on cluster mean for various characters revealed wide differences in mean value under study. The cluster III had two genotypes namely, G11 (PB 7), G20 (Parbhani Kranti) with early flowering, minimum number of nodes to first fruit, early days to first fruit harvest, maximum number of fruits per plant, maximum fruit girth and fruit yield per plant, while they have intermediate plant height. The cluster X had only one genotype G 31 (Punjab Padmini) with maximum number of branches per plant, fruit weight, fruit length, more number of fruits, high fruit girth and fruit yield per plant. The clusters VI, IV and VIII have recorded the lowest cluster mean fruit yield per plant (Table 42).

Based on the *per se* performance of different fruit yield component traits, different clusters of D^2 analysis (diversity) and cluster mean, seven genotypes were selected. They are *viz.,* G9 (Varsha Uphar), G12 (Pusa A_4), G20 (Parbani Kranti), G26 (PB 266), G31 (Punjab Padmini), G32 (Gujarat Bhindi 1) and G35 (EMS 8) from the clusters II, V, III, VII, X, I and IX respectively for further evaluation. These seven genotypes were utilized as the parents for the hybridization programme to evolve high yielding hybrids.

The genotypes *viz.,* G9, G12 and G20 which recorded maximum fruit yield per plant from different clusters also exhibited significant and favourable *per se* performance for the traits, days to first flowering, days to first fruit harvest, number of fruits per plant, fruit weight and fruit length. The genotype *viz.,* G26 had significant *per se* performance with fruit weight, plant height and fruit yield per plant from the cluster VII. Similarly, the genotypes *viz.,* G31 and G32 were also showed significant *per se* performance for the traits fruit weight, fruit length and fruit yield per plant from the cluster X and I, respectively. The genotype *viz.,* G35 registered maximum fruit yield per plant in the cluster IX coupled with significant mean value for the traits days to first flowering, days to first fruit harvest, fruit weight and plant height.

Table 40. Composition of D^2 clusters for 35 bhendi genotypes

Cluster No.	Number of genotypes	Genotypes
I	2	G29 (CO 1), G32 (Gujarat Bhindi 1)
II	12	G1 (OKH 666), G2 (NOH 303), G3 (Arka Anamika), G4 (S 51), G5 (Lakshmi), G6 (Ankur 40), G7 (DOV 1), G8 (Hissar Unnat), G9 (Varsha Uphar), G10 (Arya 351), G19 (Kamini), G33 (Har-bhajan).
III	2	G11 (PB 7), G20 (Parbhani Kranti).
IV	2	G24 (MBORH 311), G25 (DSN 1).
V	6	G12(PusaA_4),G13(Indol-03-01),G14(ArkaAbhay),G15(DOV2), G28 (Pusa Makhmali) ,G34 (Selection 2).
VI	2	G22 (DSU 1), G30 (MDU 1).
VII	2	G23 (P 7), G26 (PB 266).
VIII	4	G24 (MBORH 93), G17 (AOL-03-01), G18 (OKH 333), G21 (Bakra).
IX	2	G27 (Pusa Sawani), G35 (EMS 8).
X	1	G31 (Punjab Padmini).

Table 41. Intra-cluster and Inter-cluster distances (D^2 values) of bhendi genotypes

Clusters	I	II	III	IV	V	VI	VII	VIII	IX	X
I	25.67 (5.07)	885.36 (29.96)	499.39 (22.35)	1775.67 (42.14)	1082.62 (32.90)	970.77 (31.16)	2852.52 (53.41)	801.81 (28.32)	1193.34 (35.55)	959.44 (30.98)
II		779.64 (27.92)	1797.60 (42.40)	1081.18 (32.88)	785.43 (28.03)	637.53 (25.25)	1523.58 (39.43)	1232.76 (35.11)	769.61 (27.14)	2697.35 (51.94)
III			79.52 (8.92)	3715.55 (60.96)	2242.07 (47.35)	2122.02 (46.07)	4790.34 (69.21)	1005.30 (31.72)	2700.60 (51.97)	298.63 (17.28)
IV				80.93 (9.00)	798.57 (28.26)	570.65 (23.89)	352.44 (18.77)	2731.91 (52.26)	285.46 (16.90)	5108.76 (71.48)
V					897.42 (29.96)	598.26 (24.46)	1201.82 (34.67)	1611.01 (40.14)	615.82 (24.82)	3276.59 (57.24)
VI						174.86 (13.22)	1005.55 (31.71)	1337.95 (36.58)	516.41 (22.73)	3402.49 (58.33)
VII							222.05 (14.90)	3559.43 (59.66)	643.61 (25.36)	6455.54 (80.35)
VIII								700.22 (26.46)	2051.01 (45.29)	1489.64 (38.60)
IX									529.74 (23.02)	3848.67 (62.04)
X										0.00

Table 42. Cluster mean performance of bhendi genotypes for various characters

Traits / Clusters	Days to first flowering	Number of nodes to first fruit	Days to first fruit harvest	Number of branches per plant	Number of fruits per plant	Fruit weight (g)	Fruit length (cm)	Fruit girth (cm)	Plant height (cm)	Fruit yield per plant (g)
I	44.28	3.27	53.30	2.70	18.78	15.48	16.08	5.83	107.36	291.12
II	45.12	3.69	53.34	2.62	18.91	13.54	13.16	5.56	110.12	254.95
III	38.52	2.93	47.17	2.60	25.00	15.33	17.43	6.38	136.04	383.53
IV	45.38	4.12	54.70	2.27	13.47	14.62	11.27	5.10	69.25	196.88
V	46.22	3.08	54.10	2.37	18.61	13.79	12.71	5.59	100.06	260.73
VI	38.87	3.90	48.10	3.08	14.63	11.65	13.08	5.32	85.89	170.40
VII	42.35	3.92	52.37	3.27	18.30	14.23	9.20	4.88	83.37	261.92
VIII	44.11	4.26	52.95	3.13	20.50	10.66	15.72	5.80	126.08	216.25
IX	47.27	4.12	54.53	2.47	17.35	15.09	12.02	5.33	88.33	262.05
X	43.40	4.27	53.60	3.77	21.80	17.38	18.40	6.17	160.97	378.67

7

Genetic Diversity in Coconut [*Cocos nucifera* (L.)]

Thirugnanakumar, S. and Alice Antony

The Coconut palm is a monotypic species of the family Arecaceae, under the tribe Cocoideae. It is considered as an important commercial crop in more than 18 countries in the world. The palm provides food, drink, shelter and also supplies raw materials for a large number of industries. Every part of the palm is utilized for some economic purpose or another. Hence, it is referred to as 'Tree of Wealth" or the "Tree of Life" (Satyabalan, 1993). It is native of South East Asia. Genetic diversity has become one of the keywords of scientists who are concerned about the sustainable management of different tree crops. Genetic diversity plays an important role in tree breeding, because hybrids between populations of diverse origin, generally display a greater heterosis than between closely related parents (Zobel and Talbert, 1986). The present study was formulated to identify divergent parents for future breeding programme.

Two hundred and fifty genotypes of coconut cultivar Chawghat green dwarf, distributed in the two districts of Kerala *viz.*, Kottayam and Pathanamthitta were evaluated for nut yield per plant during June-July, 2000. Seventy five genotypes, which had single tree nut yield of more than population mean ± 2 SE were selected as per the method outlined by Ledig (1974). These 75 genotypes were observed for 10 nut and endosperm characters in two seasons (years), during June-July 2000 and 2001.

Each tree was considered as a single genotype. Thirty nuts from a single tree were collected and divided into three replications (Chauhan and Kanwar, 2001; Thirugnanakumar, 2006a, b). Observations were recorded on 10 nut and endosperm characters *viz.,* 1) oil content (%), cold percolation method; (Kartha and Sethi, 1957), 2) weight of unhusked nut (g), 3) weight of husked nut (g), 4) diameter (polar) of the husked nut (cm), 5) diameter (equatorial) of the husked nut (cm), 6) thickness of husk (cm), 7) husk: nut ratio, 8) thickness of meat (mm), 9) weight of kernel (g) and 10) weight of copra (g). The mean values so obtained were subjected to statistical analyses, in Randomized Block Design, as suggested by Chauhan and Kanwar (2001) and Thirugnanakumar *et al.* (2006a, b). The D^2 estimates were made following Mahalanobis (1936), as described by Rao (1952). Clusters were prepared following Tocher (Rao, 1952).

Seventy five genotypes were grouped into as many as 14 clusters in the first season. The number of genotypes included under various clusters ranged from two to nine (clusters II and III, respectively). The genotypes that were grouped in different clusters along with their geographical origins are furnished in Table 43. During the second season, the 75 genotypes were grouped into as many as 13 clusters (Table 44). Cluster III composed of only one genotype whereas the cluster VI comprised of as many as nine genotypes. In the pooled analysis, the seventy five genotypes were grouped into as many as 15 genotypes. Clusters VIII and X each comprised of two genotypes and the cluster IV encompassed of as many as 10 genotypes (Table 45).

It is highly desirable that the extent of genetic divergence between populations reflected through any analysis should be comparatively stable over seasons and environments, to be of utility to tree breeders. The differential grouping of different genotypes of varying origins in different seasons necessitated the need for studying the divergence in varying seasons. The question is how to select parents based on genetic divergence in varying seasons?

One may select the parents on the basis of divergence exhibited in the richest and most productive season. Because, if provides the opportunity for the fullest expression of the genetic potential of a genotype. The relative ranking of the two seasons of the present investigation revealed that the second season was best for oil content. Hence, using this criterion, the divergent genotypes in the second season may be selected from different clusters, having high inter-cluster distance. Accordingly, the following clusters are suggested for future breeding programme. The inter-cluster distance was maximum between IV and X clusters followed by IV and IX clusters (Table 46). Cluster IV exhibited maximum mean for weight of husked nut, weight of kernel and weight of copra. Cluster X exhibited high mean for husk: nut ratio (Table 46). The genotypes grouped in these clusters may be crossed to evolve high yielding lines.

Selection of parents can also be made on the basis of divergence, which is consistent over the two seasons. This can be taken as a reliable indication of genetic divergence. Utilizing this criterion, among the genotypes studied, the following genotypes were found to possess consistent divergence in all the two seasons studied. The genotypes *viz.*, G 5 and G 37 were consistently grouped in the clusters IV and VI, respectively. Crosses could well be effected among these genotypes to evolve genotypes with higher heterosis.

One may also argue that the divergence expressed in pooled analysis may be a reliable estimate and therefore should be used for selecting the parents. In pooled analysis, the 75 genotypes were grouped into as many as 15 clusters. The mean inter-cluster distance value was about 32.02 (mean divergence value was derived by dividing the total divergence values by the number of clusters, Table 47). The inter cluster distance between the clusters XI and XV was nearer to this mean divergence value. These genotypes could well be crossed, to evolve hybrids with high heterosis.

It is highly interesting to note that the genotypes which were suggested for hybridization programme based on the aforementioned three selection criteria were entirely different. The genotypes suggested based on the performance in the richest season is influenced by seasonal effect and hence it is biased. Similarly, the divergence pattern exhibited in the pooled analysis is also biased. Because, measures of divergence estimated in different seasons may cancel with each other in the pooled analysis. Thus, pooled analysis is likely to under estimate the divergence. Hence, selection of genotypes based on the consistency in the divergence pattern over seasons may hold good in the identification of divergent parents.

Table 43. Composition of D^2 clusters and their regional distribution - season I

Cluster No.	Total number of genotypes in the cluster	Regional distribution of genotypes
Cluster I	8	G 1-Chengalam; G 11-Thirumoolapuram; G 19, G 20-Vengal; G 21-Azhiyidathuchira; G 51-Vadayar; G 55, G 60-Niranam
Cluster II	2	G 54-Kadapra; G 58-Mannar
Cluster III	9	G 3-Puthanangadi; G 16-Paliyekkara; G 23-Kallissery; G 52-Vadayar; G 65-Niranam; G 18, G 61-Kadapra; G 70-Kottayam; G 71-Thazhathangadi
Cluster IV	4	G 5-Puthanangadi; G 17-Podiyadi; G 53-Niranam; G 68-Mannar
Cluster V	6	G 4-Kummanam; G 8, G 14, G 15-Kavumbhagom; G 12-Vengal; G 36-Thalayolapparampu
Cluster VI	6	G 7-Vengal; G 22-Kavumbhagom; G 30-Vadayar; G 37-Vallakom; G 63, G 67-Niranam
Cluster VII	2	G 6-Mannankarachira; G 13-Kavumbhagom
Cluster VIII	8	G 2-Puthanangadi; G 24-Kallissery; G 66-Niranam; G 69-Kadapra; G 74-Kiliroor; G 73, G 75-Kummanam; G 25-Chengannoor
Cluster IX	7	G 9-Valanjavattom; G 10, G 57-Kadapra; G 26, G 27, G 28-Vallakom; G 40-Vaikkom
Cluster X	2	G 41-Vaikkom; G 42-Vadayar
Cluster XI	3	G 29-Vadayar; G 32-Vaikkom; G 47-Vaikkom
Cluster XII	7	G 31, G 33, G 34-Vaikkom; G 38-Thalayolapparampu; G 39-Vallakom G 56-Kadapra; G 72-Kummanam
Cluster XIII	6	G 44, G 50-Vaikkom; G 62, G 64-Niranam; G 46-Thalayolapparampu; G 59-Mannar
Cluster XIV	5	G 35, G 43-Vaikkom; G 48, G 49-Vallakom; G 45-Thalayolapparampu

Table 44. Composition of D² clusters and their regional distribution - season II

Cluster No.	Total number of genotypes in the cluster	Regional distribution of genotypes
Cluster I	5	G 3-Puthanangadi; G 7-Vengal; G 18-Kadapra; G 65, G 66-Niranam;
Cluster II	4	G 15-Kavumbhago; G 32-Vaikkom; G 30-Vadayar; G 48-Vallakom
Cluster III	1	G 73-Kummanam
Cluster IV	7	G 5-Puthanangadi; G 54; G 57-Niranam; G 11-Thirumoolapuram G 10-Kadapra; G 44-Vaikkom; G 20-Vengal
Cluster V	7	G 8-Kavumbhagom; G 62, G 55-Niranam; G 17-Podiyadi; G 71-Thazhathangadi; G 19-Vengal; G 51-Vadayar
Cluster VI	9	G 29, G 42-Vadayar; G 50, G 43, G 35-Vaikkom; G 49-Vallakom; G 58-Mannar; G 37, G 45-Thalayolapparampu
Cluster VII	6	G 38, G 46-Thalayolapparampu; G 72-Kummanam; G 1-Chengalam; G 64-Niranam; G 39-Vallakom
Cluster VIII	5	G 4-Kummanam; G 21-Azhiyidathuchira G 52-Vadayar; G 36-Thalayolapparampu G 9-Valanjavattom
Cluster IX	8	G 34, G 40, G 41-Vaikkom; G 56-Kadapra G 59-Mannar; G 2-Puthanangadi; G 63-Niranam G 14-Kavumbhagom
Cluster X	8	G 13-Kavumbhagom; G 74-Kiliroor G 23, G 24-Kallissery; G 67-Niranam G 6-Mannankarachira; G 25-Chengannoor G 75-Kummanam
Cluster XI	3	G 22-Kavumbhagom; G 33-Vaikkom G 12-Vengal
Cluster XII	7	G 26, G 27, G 28, G 47-Vallakom G 70-Kottayam; G 31- Vaikkom; G 69-Kadapra
Cluster XIII	5	G 16-Paliyakkara G 53, G 60, G 61, G 68-Niranam

Table 45. Composition of D^2 clusters and their regional distribution - seasons pooled

Cluster No.	Total number of genotypes in the cluster	Regional distribution to genotypes
Cluster I	5	G 1-Chengalam; G 11-Thirumoolapuram G 21-Azhiyidathuchira; G 52-Vadayar G 57-Niranam
Cluster II	5	G 9-Valanjavattom; G 10-Kadapra; G 38-Thalayolapparampu; G 39-Vallakom G 70-Kottayam
Cluster III	6	G 44-Vaikkom; G 46-Thalayolapparampu G 58-Mannar; G 64, G 62-Niranam G 71-Thazhathangadi
Cluster IV	10	G 3, G 5-Puthanangadi; G 8, G 15-Kavumbhagom G 7-Vengal; G 17-Podiyadi; G 18-Kadapra G 30-Vadayar; G 36-Thalayolapparampu; G 66-Niranam
Cluster V	6	G 19, G 20-Vengal; G 51-Vadayar G 54, G 55, G 65-Niranam;
Cluster VI	7	G 4-Kummanam; G 14, G 22-Kavumbhagom; G 34-Vaikkom; G 56-Kadapra; G 37-Thalayolapparampu; G 63-Niranam
Cluster VII	6	G 41, G 40-Vaikkom; G 42-Vadayar G 13-Kavumbhagom; G 67-Niranam G 67-Mannankarachira
Cluster VIII	2	G 23-Kallissery; G 73-Kummanam
Cluster IX	4	G 24-Kallissery; G 2-Puthanangadi G 74-Kiliroor; G 69-Kadapra
Cluster X	2	G 25-Chengannoor; G 75-Kummanam
Cluster XI	6	G 12-Vengal; G 28, G 27, G 26-Vallakom; G 31, G 33-Vaikkom
Cluster XII	3	G 29-Vadayar; G 47-Vallakom G 32-Vaikkom
Cluster XIII	5	G 35, G 43, G 50-Vaikkom; G 59-Mannar; G 72-Kummanam
Cluster XIV	3	G 45-Thalayolapparampu; G 49, G 48-Vallakom
Cluster XV	5	G 16-Paliyakkara G 53, G 60, G 61, G 68-Niranam

Table 46. Inter and intra (Diagonal) cluster average of D values and extent of Diversity among clusters – season II

Cluster Number	Cluster I	Cluster II	Cluster III	Cluster IV	Cluster V	Cluster VI	Cluster VII	Cluster VIII	Cluster IX	Cluster X	Cluster XI	Cluster XII	Cluster XIII
Cluster I	3.13	3.50	5.24	3.63	3.67	4.24	4.30	3.83	4.33	4.40	5.55	6.07	5.21
Cluster II		2.46	6.84	3.73	4.13	3.75	3.90	3.09	3.95	4.32	4.51	5.64	6.10
Cluster III			0.00	5.74	6.00	7.10	6.40	6.07	6.41	5.72	7.79	7.45	7.52
Cluster IV				1.78	2.79	2.93	2.42	2.34	2.75	3.72	4.69	4.09	7.05
Cluster V					2.89	3.61	3.36	3.44	4.07	4.44	5.46	5.25	6.21
Cluster VI						2.68	2.76	3.19	3.22	4.49	4.13	4.05	7.43
Cluster VII							1.79	2.40	3.01	3.95	3.71	3.15	7.82
Cluster VIII								1.47	2.51	2.99	3.70	3.87	7.28
Cluster IX									2.31	3.35	3.84	3.67	7.86
Cluster X										2.93	4.24	4.78	7.35
Cluster XI											3.04	3.83	8.71
Cluster XII												2.78	9.76
Cluster XIII													3.96

Table 47. Inter and intra (Diagonal) cluster average of D values and extent of Diversity among clusters - seasons pooled

Cluster Number	Cluster I	Cluster II	Cluster III	Cluster IV	Cluster V	Cluster VI	Cluster VII	Cluster VIII	Cluster IX	Cluster X	Cluster XI	Cluster XII	Cluster XIII	Cluster XIV	Cluster XV
Cluster I	2.10	2.47	2.89	3.68	2.82	3.30	4.57	3.50	3.36	3.89	4.06	4.61	3.92	5.17	6.36
Cluster II		1.78	2.89	4.00	3.49	2.73	4.57	3.92	3.07	4.25	3.04	3.48	3.55	5.16	6.91
Cluster III			2.74	3.80	3.24	3.53	4.78	4.05	4.11	5.12	4.04	4.33	3.16	4.43	6.27
Cluster IV				2.98	3.37	3.59	4.60	4.02	4.71	5.24	5.17	5.22	4.03	4.60	4.72
Cluster V					2.33	3.99	5.16	4.16	4.61	5.21	5.33	5.38	4.25	5.17	5.29
Cluster VI						2.03	3.63	3.72	2.87	4.21	3.33	4.17	3.11	4.57	6.51
Cluster VII							4.49	4.31	4.09	5.05	4.79	5.77	4.10	4.95	7.05
Cluster VIII								3.15	3.26	3.62	4.95	5.94	4.49	5.65	5.96
Cluster IX									2.06	2.88	3.70	5.23	4.24	5.95	7.49
Cluster X										0.72	4.83	6.36	5.69	7.03	7.77
Cluster XI											2.47	3.30	3.66	5.03	8.35
Cluster XII												3.05	4.24	5.12	8.06
Cluster XIII													2.11	3.06	6.82
Cluster XIV														2.84	6.99
Cluster XV															3.60

8

Genetic Diversity in Pomegranate [*Punica granatum* (L.)]

Manivannan, K.

Pomegranate is one of the popular fruit crops of tropical and sub tropical in India and enjoys high reputation for its dietetic and medicinal properties. The fruit is native of Iran and is extensively cultivated in Mediterranean countries like Spain, Morocco, Egypt, Iran, Afghanistan. It is also grown to a lesser extent in Myanmar, China, Japan, India and U.S.A.

The area under pomegranate cultivation has increased several folds in the recent past after the introduction of soft seeded varieties. Out of 30,000 hectares under this crop in India, Maharashtra state alone grows over eighty per cent of area with an average production of 8000 kg ha^{-1} (Mote *et al.*, 1992). The export of pomegranate fruits was 1790.3 tonnes in 1991-92 valued at rupees 215.3 lakhs which was more than two fold over the previous year's export earnings (Chadha, 1994). The utility of the crop is gaining importance and at present new orchards of pomegranate are being planted on large acreage. Because of its versatile adaptability, hard nature, low maintenance cost, steady and high yields, fine table and therapeutic values, better keeping quality and very good export potential, there is greater scope for extending its area and increasing the production especially in the hot and semi arid regions of India.

In pomegranate, market preference is for the attractiveness of the fruit which are exhibited through better size, colour with bold and deep coloured arils besides possessing high juice-content. However consumers prefers the softness of the arils with low acidity. While, there are number of cultivated types available with these desirable characters, there are hardly cultivar having a combination of all or atleast a majority of them. Pomegranate improvement work, therefore has to be oriented towards the production of soft seeded type with other desirable characters (Phadnis, 1974).

Combination or recombination of certain desirable characters are only possible through hybridization. Broadening the genetic base of the existing population is also possible by hybridization which has proved its superiority over selection. Further multiplication of the true to type progenies could be achieved easily through clonal propagation.

Lack of variation in chromosome number of morphological characters among the varieties would necessiate initiating hybridization for pomegranate improvement. Combination of characters like profuse flowering, larger fruit size and greater edible portion of fruits would be desirable. The availability of promising features in different stocks, the ease of hybridization and seed propagation facilitate the process and it would be worthy of the time and efforts spent on such programme according to Raman *et al.* (1963).

Pomegranate has received comparatively little attention in crop improvement compared to other crops. Being an often cross pollinated crop, it is suited for heterosis breeding. A precise knowledge of genetic divergence, combining ability and gene action responsible for inheritance of desirable characters such as larger fruit size, high yield, better fruit set, soft seededness, bold arils with deep colour are the basic pre-requisites for conducting a successful hybridization programme in pomegranate.

The exploitation of "Hybrid vigour" in pomegranate is very meager. Information on hybrid vigour would enable to bring promising hybrids with desirable characters and to increase the fruit yields as reported in mango by Naik *et al.* (1959).

An important objective in any crop improvement programme is to enhance genetic potential for fruit yield. Genetic divergence considered as one of the criteria for selection of the parents to generate crosses which segregate in later generations into genotypes, transgressing the performance of the better parents. The quantitative trait *viz.,* yield is often influenced by the environment. Therefore the estimates of genetic and environmental variation are to be assessed. Further, a knowledge of heritability is needed to reflect the possibility and extend to which the improvement is possible through selection.

Grafius (1956) in proposing a genetic concept of yield, suggested that yield might be more effectively increased by selecting one or more yield components. According to Falconer (1960), yield improvement by yield component selection should be superior to selection for yield *per se.* when the component character has a higher heritability than yield and when the correlation between the two traits is of greater importance in designing either a single-trait-selection or index selection to improve the yield.

The great range of genetic variability available among the varieties, the wide range of adaptation of this crop and its medicinal properties are some of the important aspects calling for intense efforts to improve this crop. Efforts are to be directed to identify the potential parents through more refined biometrical techniques to classify the parents and hybrids in terms of their ability to combine in hybrid combinations. The genetic studies leading to the nature of gene action is also scanty in this crop. It is also necessary to collect more information on the inheritance pattern of economic characters so as to include parents with desirable attributes in the breeding programme. Therefore, an investigation on the hybridization and evaluation of F_1 hybrids along with the parents was taken up.

The present investigation on pomegranate (*Punica granatum* L.) was carried out by Manivannan (1995) at the Horticultural Research Station, Tamil Nadu Agricultural University, Yercaud during 1991 to 1995. The location of the present study is situated at 1350 m above M.S.L. in the Shevrayan hills ranges of South India. The average annual rainfall is 1200 mm. The maximum temperature ranges from 20°C to 32°C and the minimum from 8°C to 18°C. The soil type is lateritic loam with a pH of 5.5 to 6.5.

Six promising cultivars of pomegranate were selected from the germplasm collection and utilized for the study. The parent materials involved in crossing were selected based on their performance in relevance to the following characters.

S. No.	Cultivars	Characters
1	Ganesh	Soft seeded, high yielding, high fruit weight
2	Jyothi	More grain/peel ratio, tolerance to fruit borer
3	Yercaud-1	Vigorous growth, soft seeded, high reducing sugar
4	Kabul	Bold arils, high TSS
5	Coimbatore	Early bearing, high ascorbic acid
6	Bedana	More number of hermaphodite flowers with high setting percentage

All possible one-way cross combinations were effected with a view to have improvement in yield, fruit, size, soft seededness and quality. Successful fruit set was obtained only in eight crosses and the hybrid progenies were raised for evaluation. The shoot tip cuttings of the parent materials were rooted in the mist chamber and utilized for comparison in the present study.

Emasculation of flowers was done on the previous day of anthesis, by manual removal of the peals. The flowers were then bagged to avoid open pollination.

During the next day, the stigma of the emasculated flower was dusted with the pollen collected from the flowers of male parents involved in the present study and the pollinated flowers were bagged immediately.

The crossing programme was carried out between August 1991 and October 1991. The crossed and fully matured were harvested after 140 days from fruitset and the fresh seeds collected were sown in seed pans after removing the pulp. The seedlings were then transferred to the polythene bags after germination. They were transplanted after ninety days in the main field with a spacing of 3.0 m × 3.0 m in June 1992. Ten rooted cuttings in each of the six parents were planted along with a total of 953 hybrids of comparison. The crop was maintained adopting all the improved package of practices uniformly to all the plants. The plants started flowering during October 1994.

The following observations in respect of vegetative growth, flowering and fruiting were recoded during 1994-95 on individual plant basis.

Cluster analysis with K-means algorithm (Hartigan, 1975) was used as a data reduction technique in the identification of superior hybrids among all the 365 hybrids, treated with a total eight yield traits. Hybrids with a highest Euclidean distance with the highest cluster mean values were selected from the seven clusters. The cluster analysis was done in a PC-AT computer with CoRes software of Pascal International Software Solutions (1993).

Clustering procedure was performed for eight fruits *viz.*, plant height, plant spread (East-West), number of hermaphrodite flowers per plant, number of fruits per plant, fruit yield, single fruit weight, fruit length and breadth, using K-means algorithm, first for all parents and hybrids together (425 plants), then hybrids of each cross with their respective parents, keeping the number of clusters as seven.

The clustering procedure was performed for all the four hundred and twenty five plants (365 hybrids and 60 parental plants), grouping into seven cluster. The first cluster had the highest number of 106 plants which was followed by the fifth cluster with 92 plants and the lowest of twelve plants in the fourth cluster. The remaining sixth, seventh, second and third clusters had 84, 59, 44 and 28 plants respectively.

The mean performed of the seven clusters showed that the plant height was the highest (250.03 cm) in the third cluster followed by the fourth cluster (237.86 cm) and the lowest (106.99 cm) in the first cluster. The spread of the plant in the East-West direction was the widest (195.09 cm) in the fourth cluster followed by the third cluster (160.18 cm). The plant spread was very narrow (90.61 cm) in the first cluster.

The mean number of hermaphrodite flowers produced per plant was maximum (30.67) in the fourth cluster followed by the third one (29.49). The minimum number of 17.69 per plant flowers was recorded in the first cluster. In respect of number of fruits produced per plant, the third cluster ranked first with 17.51 followed by 13.08 in fourth cluster. The number of fruits per plant was the least (7.67) in the first cluster.

Regarding the yield of fruits per plant, highest yield of 4.21 kg was registered by the cluster IV followed by the cluster III with 3.44 kg per plant. The lowest yield of 1.41 kg was recorded in the cluster VII. For single fruit weight, the cluster IV had the highest value of 324.33 g/fruit closely followed by the cluster II (292. 60 g/fruit) and the lowest of 143.41 g/fruit was in the cluster VIII.

The size of the fruit, in respect of length and breadth was maximum (8.08 and 8.00 cm) in the cluster IV followed by the cluster II (7.31-7.45 cm) and the minimum fruit size (6.22-6.36 cm) was observed in the cluster VII.

For the overall performance of all the traits, the cluster IV was outstanding. Among the 11 members in the cluster there were only two hybrids *viz.,* GA. YCD-7 and GA.BA-18, while the rest comprised of nine plants of Ganesh and one plant of Yercaud-1. This cluster had the highest cluster mean values for fruit yield (4.21 kg/plant), fruit weight (324.33 g), number of flowers (36.67 kg/plant), plant spread (195.09 cm in E-W) and fruit size (length 8.08 cm; Breadth 8.00 cm) and it was second only to cluster III for plant height and number of fruits per plant. The highest distance metrics was exhibited by GA.BA-18 with a Euclidean distance of 27.95.

When the K-means clustering procedure was carried out with hybrids and their respective parents alone it was found that the parents had better distance metrics and cluster statistic than their hybrids. Hence hybrids with the best distance metrics and the cluster statistics alone are tabulated and the details are reported below.

Parents and hybrids of Ganesh × Jyothi

The clustering procedure was performed for all the sixty hybrids and ten plants each of Ganesh and Jyothi. The plants of the parent Ganesh congregated in the fifth and sixth clusters ranking high for all the eight characters. Whereas, the cluster II comprised of seven hybrids and seven plants of the male parent Jyothi which ranked second with respect to fruit yield (2.87 kg/plant) and single fruit weight (298.53 g). Among the seven hybrid plants in the cluster-II, the hybrid GA.JO-35 was having the highest Euclidean distance of 12.16 followed by GA.JO-12 with 12.52. The lowest Euclidean distance of 4.74 was for the hybrid GA.JO-57.

The cluster statistics revealed that the plant height ranged from 144.50 to 196.80 cm with a mean value of 177.35 cm. The mean spread of canopy in the East-West direction was 135.69 cm ranging from 110.50 to 161.50 cm. The mean number of flowers per plant was 25.36 with a mean of 10.43 fruits per plant. The fruit yield ranged from 1.90 to 3.64 kg with a mean value of 2.87 kg per plant. The mean single fruit weight was 298.53 g with the fruit size of 7.33 and 7.34 cm in length and breadth respectively.

Parents and hybrids of Ganesh × Yercaud-1

The clustering was effected for all the sixty eight hybrids and ten plants each of Ganesh and Yercaud-1. Both of the parents fell in line the clusters II and III which recorded the higher mean values for all the characters taken into account for making the clusters.

The cluster VII was constituted by the best five hybrids GA.YCD-7, 8, 35, 37 and 58. Among them, the hybrid GA.YCD-7 was having the highest Euclidean distance of 13.38 which inturn was followed by GA.YCD-38 with 9.09. The lowest Euclidean distance of 5.34 was held by GA.YCD-35. This cluster was ranked second next to cluster III for single fruit weight with a value of 298.58 g as well as fruit size (7.66-7.96 cm). It was interesting to note that the hybrid GA.YCD-7 had a higher Euclidean distance when the clustering was performed for hybrids and parents overall previously.

Parents and hybrids of Ganesh × Kabul

The clustering was effected for all the sixty eight hybrids and ten plants each of Ganesh and Kabul making seven clusters. The parent Ganesh with nine plants found its place in the cluster-II which had the highest values for all the characters except total number of fruits per plant. The cluster IV isolated itself with two hybrids plants *viz.,* GA.KA-48 and GA.KA-58 with a Euclidean distance of 12.50 and 8.53 respectively and a lone parent GA-6. This cluster had the highest value for number of fruits per plant (15.67) recording the second best yield of 3.11 kg fruit per plant.

Parent and hybrids of Ganesh × Coimbatore

Seven clusters were made out of twenty five hybrids and ten plants each of Ganesh and Coimbatore. The cluster VII ranked first for all the characters except total number of fruits per plant, in which two plants of Ganesh were seen. The second best cluster VI was having only one hybrid viz., GA.CBE-13. The mean yield was the second best among the seven clusters with 4.38 kg fruits per plant and the single fruit weight was 274.00 g.

Parents and hybrids of Ganesh × Bedana

A total of 45 hybrids and ten plants each of Ganesh and Bedana were made into seven clusters. The cluster III ranked first for most of the characters in which nine plants of Ganesh found their place. The cluster V was having the maximum plant height (243.90 cm) which was with sixteen hybrid plants in its groove. Among the hybrids in this cluster, the hybrid GA.BA-18 was having the highest Euclidean distance of 21.09 followed by GA.BA-26 with 17.67. The least value of 6.95 was scored by GA.BA-32. The plant height ranged from 209.90 to 271.20 cm with a mean of 243.90 cm. The plant spread in East-West direction ranged between 110.30 and 186.30 cm with a mean value of 144.75 cm. The mean number of flowers and fruits were 24.44 and 14.56 respectively. The fruit yield ranged from 1.87 to 4.64 kg per plant with a mean of 3.48 kg per plant. Single fruit weight had the cluster mean value of 239.00 g with the fruit size of 7.01 and 7.29 cm in length and breadth respectively.

Parents and hybrids of Jyothi × Yercaud-1

The clustering was performed for thirty eight hybrids and ten plants each of Jyothi and Yercaud-1. The cluster-1 had the highest mean value for both plant height (278.90 cm) and plant spread in East-West direction (167.64 cm)and constituted by the male parent Yercaud-1. The total number of flowers and fruits per plant were highest (31.50 and 15.50 respectively) in the cluster-VI, in which one hybrid JO.YCD-37 and three plants of Jyothi *viz.*, JO-6, JO-8 and JO-10 found their place. The cluster IV in which the remaining seven plants of Jyothi were located got the maximum mean value for fruit weight (287.49 g per fruit) and the fruit length (7.23 cm) and breadth (7.40 cm).

In the cluster VI, the Euclidean distance was the highest (13.03) for the hybrid JO.YCD-37 with the higher yield of 3.43 kg per plant and a mean fruit weight of 221.85 g per fruit.

Parents and hybrids of Jyothi × Kabul

The clustering procedure was effected for 26 hybrids and ten plants each of jyothi and Kabul. The clusters VI and VII showed better performance and it was noted that the two hybrids, viz., JO.KA-10 and JO.KA-29 excelled the parents. The lone hybrid JO.KA-10 constituted the cluster-VI which secured the highest value for the plant spread in East-West direction (198.20 cm) and the highest fruit yield of 3.54 kg per plant. For plant height, the cluster-VII had highest mean value of 190.50 cm and ranked first for the single fruit weight (337.00 g) and the size of the fruit (length 8.30 cm and breadth 3.10 cm). In this cluster only one hybrid *i.e.*, JO.KA-29 found a place.

Parents and hybrids of Yercaud-1 × Kabul

Clustering was performed for all the 35 hybrids and ten plants each of Yercaud-1 and Kabul for grouping in to seven clusters. Both the clusters II and VII were constituted by the parents *viz.*, Yercaud-1 and Kabul respectively scored the highest values for all the characters except single fruit weight and fruit size. The cluster IV comprised of nine hybrid plants which got the highest values for single fruit weight (234.32 g per fruit) and fruit size (length 7.17 cm; breadth 7.42 cm).

The widest Euclidean distance of 13.63 was registered by the hybrid YCD. KA-14 followed by YCD.KA-25 (12.06). The fruit yield ranged from 1.89 to 3.82 kg per plant with the mean value of 2.97 kg per plant.

Each hybrid of the eight crosses was studied thoroughly for the growth and flowering behaviour, yield and physico-chemical characteristics of the fruit. Out of the 365 hybrid progenies, one from each cross found superior for one or more characters was identified.

A breeder is posed with the problem of selecting a few from the mixed bag of gene pool for further evaluation and forwarding them to greater exploitation by endusers. It is daunting task to select for various traits from a voluminous assembly for genotypes. Under such circumstances, the breeder is forced to classify the gene pool first, by grouping the segregants before selection is exercised.

Primarily, classification represent one of the many possible ways in which vegetational data may be analysed. Secondly, the user must be satisfied that classifications can do something uniquely well under the circumstances in which other methods could not procedure better results. Cluster analysis is one such classification analysis, employed in the present study, using the computational method provided by Hartigan (1975).

In the present study, sixty plants of six parents and 365 hybrids from eight crosses were classified into seven clusters using K-means algorithm using the multivariate data with eight varieties viz. plant height, plant spread (East-West), number of hermaphrodite flowers per plant, number of fruits per plant, fruit yield, single fruit weight, fruit size (length and breadth). It was found that cluster IV had the best cluster statistics. This cluster had two hybrids, *viz.*, GA.BA-18 and GA.YCD-7 and ten parental plants with a cluster mean yield of 4.21 kg fruits per plant. But the plants were tall with a mean value of 237.86 cm, as this cluster contained more of parental plants of Ganesh and Yercaud-1, which were taller. This cluster had a mean yield of 13.08 fruits per plant while the highest number of fruits (17.51) were found in cluster III. But in the cluster IV the single fruit weight (324.33 g), fruit breadth (3.08 cm) and fruit length (8.0 cm) were the highest. This again, was adduced by the findings in path analysis where the number of fruits had a high negative direct effect while other traits had a positive direct effect over yield. In all, cluster IV with twelve plants could be adjudged the best among the total of 425 plants in the study.

The clustering procedure was performed for hybrids of each cross along with their respective parents to identify the superior hybrids among the different cross combinations.

In the analysis with parents and hybrids of Ganesh and Bedana, cluster V was found to be the best which had the best cluster statistics of 3.48 kg of fruit yield/plant among the hybrids, while the cv. Ganesh showed a better statistics. Interestingly the cluster V contained the plant GA.BA-18 which was found to be the best hybrid in the cluster IV of previous analysis with all the hybrids and parental plants thus adducing its supremacy.

In the analysis with parents and hybrids of Ganesh × Yercaud-1, cluster VII was found to be the best with better cluster statistics. The hybrid GA.YCD-7 was adjudged best in this cluster because it had better distance metrics and it was found along with GA.BA-18 in cluster IV of overall with all the plants of study.

Considering the above interference it could be concluded that the hybrids GA.BA-18 with the yield performance (fruit yield 4.64 kg/plant; number of adjudged the best. The hybrid GA.YCD-7 produced larger fruits (312.0 g) of fewer number (12.0) as compared to smaller fruits (290.0 g) of larger number (16.0) in the case of GA.BA-18. However, in terms of quality, GA.YCD-7 had better attributes (seed hardness of 1.20 kg/cm^2, TSS of 17.8% less acidity of 0.46% and reducing sugar 13.80%) than GA.BA-18 (seed hardness of 2.16 kg/cm^2, TSS of 17.2%, acidity of 0.65% and reducing sugar 11.4%). The hybrid GA.YCD-7 for the best quality with better yields and GA.BA-18 for the best yields with moderate quality along with other identified ones from the remaining six of the eight cross combinations can be further multiplied vegetatively for evaluation with more population or crosses with superior varieties in future hybridization programmes.

Table 48. Cluster analysis for parents and hybrids of pomegranate (Cluster IV)

Hybrid/Parent	Euclidean distance	Parent	Euclidean distance
GA.YCD-7	16.71	GA.5	13.17
GA.BA-18	27.95	GA.7	8.75
GA.1	6.41	GA.8	7.55
GA.2	10.95	GA.9	24.62
GA.3	14.68	GA.10	21.69
GA.4	7.68	YCD.3	24.09

Cluster statistics

Characters	Minimum	Mean	Maximum	S.D
Plant height (cm)	190.50	237.86	271.20	18.15
Plant spread (cm) (East-West)	132.40	195.80	216.80	29.94
Number of hermaphrodite flowers per plant	24.00	36.67	42.00	4.85
Number of fruits per plant	10.00	13.08	16.00	1.89
Fruit yield (g)	3.13	4.21	4.76	0.54
Single fruit weight (g)	286.00	324.33	390.00	33.07
Fruit length (cm)	6.40	8.08	9.20	0.88
Fruit breadth (cm)	7.00	8.00	9.00	0.65

Table 49. Cluster analysis for parents and hybrids of Ganesh × Jyothi (Cluster II)

Hybrid/Parent	Euclidean distance	Parent	Euclidean distance
GA.JO-12	12.52	JO-1	12.53
GA.JO-29	10.63	JO-2	3.59
GA.JO-35	12.61	JO-3	15.21
GA.JO-51	11.49	JO-4	6.85
GA.JO-55	8.27	JO-5	14.87
GA.JO-56	8.72	JO-7	7.80
GA.JO-57	4.74	JO-9	8.00

Cluster statistics

Characters	Minimum	Mean	Maximum	S.D
Plant height (cm)	144.50	177.35	196.80	15.92
Plant spread (cm) (East-West)	110.50	135.69	161.50	13.33
Number of hermaphrodite flowers per plant	18.00	25.36	30.00	3.18
Number of fruits per plant	6.00	10.43	15.00	2.58
Yield (g)	1.90	2.87	3.64	0.47
Single fruit weight (g)	260.00	298.53	337.00	20.52
Fruit length (cm)	6.40	7.33	8.30	0.42
Fruit breadth (cm)	6.40	7.34	8.10	0.45

Table 50. Cluster analysis for parents and hybrids of Ganesh × Yercaud-1 (Cluster VII)

Hybrid	Euclidean distance
GA.YCD-7	13.38
GA.YCD-8	6.42
GA.YCD-35	5.34
GA.YCD-37	8.78
GA.YCD-58	9.09

Cluster statistics

Characters	Minimum	Mean	Maximum	S.D
Plant height (cm)	166.20	185.46	217.90	20.35
Plant spread (cm) (East-West)	130.30	141.74	154.30	7.63
Number of hermaphrodite flowers per plant	18.00	27.00	36.00	6.39
Number of fruits per plant	6.00	9.20	13.00	2.32
Yield (g)	1.74	2.76	3.92	0.73
Single fruit weight (g)	281.00	298.58	312.60	11.58
Fruit length (cm)	6.90	7.66	8.20	0.43
Fruit breadth (cm)	7.30	7.96	8.60	0.43

Table 51. Cluster analysis for parents and hybrids of Ganesh × Kabul Cluster IV

Hybrid/Parent	Euclidean distance
GA.KA-48	12.50
GA.KA-58	8.53
GA-6	17.10

Cluster statistics

Characters	Minimum	Mean	Maximum	S.D
Plant height (cm)	167.30	206.23	237.00	21.95
Plant spread (cm) (East-West)	140.60	157.77	190.50	23.16
Number of hermaphrodite flowers per plant	21.00	28.00	42.00	9.90
Number of fruits per plant	14.00	15.67	18.00	1.70
Yield (g)	2.59	3.11	3.96	0.60
Single fruit weight (g)	185.00	197.13	220.40	16.46
Fruit length (cm)	6.20	6.70	7.00	0.36
Fruit breadth (cm)	6.80	6.93	7.20	0.19

Table 52. Cluster analysis for parents and hybrids of Ganesh × Coimbatore (Cluster VI)

Hybrid	GA.CBE-13

Cluster statistics

Characters	GA.CBE-13
Plant height (cm)	216.20
Plant spread (cm) (East-West)	107.10
Number of hermaphrodite flowers per plant	37.00
Number of fruits per plant	16.00
Yield (g)	4.38
Single fruit weight (g)	274.00
Fruit length (cm)	7.60
Fruit breadth (cm)	7.80

Table 53. Cluster analysis for parents and hybrids of Ganesh × Bedana (Cluster V)

Hybrid/Parent	Euclidean distance	Parent	Euclidean distance
GA.BA-7	14.77	GA.BA-26	17.67
GA.BA-9	10.04	GA.BA-28	12.11
GA.BA-10	10.46	GA.BA-29	15.25
GA.BA-11	12.61	GA.BA-31	10.76
GA.BA-12	8.64	GA.BA-32	6.95
GA.BA-18	21.09	GA.BA-38	13.97
GA.BA-19	7.35	GA.BA-40	13.12
GA.BA-20	7.88	GA.BA-42	11.79

Cluster statistics

Characters	Minimum	Mean	Maximum	S.D
Plant height (cm)	209.90	243.90	271.20	19.46
Plant spread (cm) (East-West)	110.30	144.75	186.30	21.89
Number of hermaphrodite flowers per plant	21.00	24.44	32.00	2.94
Number of fruits per plant	9.00	14.56	20.00	3.28
Yield (g)	1.87	3.48	4.64	0.82
Single fruit weight (g)	208.00	239.00	290.00	20.36
Fruit length (cm)	6.00	7.01	7.40	0.29
Fruit breadth (cm)	6.20	7.29	7.90	0.31

Table 54. Cluster analysis for parents and hybrids of Jyothi × Yercaud-1 (Cluster IV)

Hybrid/Parent	Euclidean distance
JO.YCD-37	13.03
JO-6	9.01
JO-8	5.37
JO-10	4.91

Cluster statistics

Characters	Minimum	Mean	Maximum	S.D
Plant height (cm)	181.20	191.10	196.80	6.14
Plant spread (cm) (East-West)	130.60	152.88	187.50	21.24
Number of hermaphrodite flowers per plant	28.00	31.50	38.00	3.84
Number of fruits per plant	14.00	15.50	17.00	1.12
Yield (g)	3.21	3.43	3.84	0.24
Single fruit weight (g)	210.40	221.85	236.40	10.16
Fruit length (cm)	6.80	7.07	7.20	0.16
Fruit breadth (cm)	7.00	7.15	7.40	0.15

Table 55. Cluster analysis for parents and hybrids of Jyothi × Kabul

Cluster VI	Cluster VII
JO.KA-10	JO.KA-29

Cluster statistics

Characters	JO.KA-10	JO.KA-29
Plant height (cm)	160.10	190.50
Plant spread (cm) (East-West)	198.20	149.60
Number of hermaphrodite flowers per plant	26.00	24.00
Number of fruits per plant	15.00	10.00
Yield (g)	3.54	3.37
Single fruit weight (g)	236.20	337.00
Fruit length (cm)	7.10	8.30
Fruit breadth (cm)	7.40	8.10

Table 56. Cluster analysis for parents and hybrids of Yercaud-1 × Kabul (Cluster IV)

Hybrid/Parent	Euclidean distance
YCD.KA-9	10.76
YCD.KA-14	13.63
YCD.KA-16	9.55
YCD.KA-22	5.79
YCD.KA-25	12.06
YCD.KA-30	10.81

Cluster statistics

Characters	Minimum	Mean	Maximum	S.D
Plant height (cm)	132.50	168.12	186.50	19.11
Plant spread (cm) (East-West)	140.50	155.77	170.80	11.90
Number of hermaphrodite flowers per plant	16.00	26.67	32.00	5.59
Number of fruits per plant	9.00	12.67	16.00	2.81
Yield (g)	1.89	2.97	3.82	0.70
Single fruit weight (g)	210.40	234.32	264.50	19.28
Fruit length (cm)	7.00	7.17	7.40	0.14
Fruit breadth (cm)	7.20	7.42	7.60	0.17

Table 57. Composition of clusters for parents and hybrids

Group/ Cluster	Parents						Hybrids								Total
	GA	JO	YCD-1	KA	CBE	BA	GAJO	GAY-CD	GAKA	GACBE	GABA	JOY-CD	JOKA	YC-DKA	
I	-	-	9	-	-	-	15	36	21	7	-	9	6	3	106
II	-	4	-	-	-	-	14	12	10	3	-	1	-	-	44
III	1	1	-	-	-	10	-	-	-	-	16	-	-	-	28
IV	9	-	1	-	-	-	-	1	-	-	1	-	-	-	12
V	-	-	-	10	-	-	6	12	12	6	11	10	6	19	92
VI	-	5	-	-	10	-	10	5	13	4	12	8	9	8	84
VII	-	-	-	-	-	-	15	2	12	5	5	10	5	5	59
Total	10	10	10	10	10	10	60	68	68	25	45	38	26	35	425

Table 58. Mean performance of parents and hybrids in each cluster for yield and attributing traits in pomegranate

Character/ cluster	Plant height (cm)	Plant spread East-West (cm)	Number of hermaphrodite flowers per plant	Number of fruits per plant	Fruit yield (kg)	Single fruit weight (g)	Fruit size	
							Length (cm)	Breadth (cm)
I	106.99	90.61	17.69	7.97	1.73	218.17	6.85	7.06
II	147.37	117.43	22.48	9.36	2.67	292.60	7.31	7.45
III	250.03	160.18	29.49	17.51	3.44	202.93	6.76	6.98
IV	237.86	195.09	36.67	13.08	4.21	324.33	8.08	8.00
V	144.21	114.21	23.05	11.57	2.08	182.46	6.55	6.73
VI	171.54	129.53	25.11	12.33	2.82	228.50	7.01	7.20
VII	108.43	91.14	20.42	9.73	1.41	143.41	6.22	6.36
Grand mean	166.63	128.31	24.99	11.65	2.62	227.49	6.97	7.11

Table 59. Mean performance of each cluster of parents and hybrids in Ganesh × Jyothi

Character/ cluster	Plant height (cm)	Plant spread East-West (cm)	Number of hermaphrodite flowers per plant	Number of fruits per plant	Fruit yield (kg)	Single fruit weight (g)	Fruit size	
							Length (cm)	Breadth (cm)
I	103.03	94.39	27.59	10.29	1.53	151.84	6.29	6.28
II	177.35	135.69	25.36	10.43	2.87	298.53	7.33	7.34
III	114.28	105.05	22.04	7.58	1.94	250.17	7.04	7.17
IV	162.98	131.63	25.00	10.36	2.22	213.83	6.78	6.94
V	241.63	211.64	39.11	13.44	4.38	328.04	8.16	8.04
VI	237.00	190.50	42.00	15.00	2.79	186.00	6.20	6.80
VII	147.55	112.90	28.75	1.25	1.41	133.15	6.30	6.20
Grand mean	169.12	140.26	29.98	11.19	2.45	223.08	6.87	6.97

Table 60. Mean performance of each cluster of parents and hybrids in Ganesh × Yercaud-1

Character/ cluster	Plant height (cm)	Plant spread East-West (cm)	Number of hermaphrodite flowers per plant	Number of fruits per plant	Fruit yield (kg)	Single fruit weight (g)	Fruit size	
							Length (cm)	Breadth (cm)
I	128.43	104.12	19.57	7.29	2.01	272.66	7.24	7.41
II	275.09	169.72	27.82	15.36	3.02	197.22	6.70	6.89
III	241.63	211.64	39.11	13.44	4.38	328.04	8.16	8.04
IV	105.56	87.15	15.74	6.06	1.32	213.27	6.80	7.00
V	129.85	102.20	17.00	6.00	0.81	136.25	6.35	7.00
VI	149.74	120.01	24.06	7.94	1.56	197.01	6.70	6.81
VII	185.46	141.74	27.00	9.20	2.76	298.58	7.66	7.96
Grand mean	173.68	133.80	24.33	9.33	2.28	234.72	7.09	7.30

Table 61. Mean performance of each cluster of parents and hybrids in Ganesh × Kabul

Character/ cluster	Plant height (cm)	Plant spread East-West (cm)	Number of hermaphrodite flowers per plant	Number of fruits per plant	Fruit yield (kg)	Single fruit weight (g)	Fruit size	
							Length (cm)	Breadth (cm)
I	157.27	127.37	23.40	11.87	2.86	240.23	6.95	7.16
II	241.63	211.64	39.11	13.44	4.38	328.04	8.16	8.04
III	146.43	109.45	28.25	13.79	2.31	168.08	6.43	6.58
IV	206.23	157.77	28.00	15.67	3.11	197.13	6.70	6.93
V	88.20	85.10	20.00	6.00	0.46	78.00	7.40	7.60
VI	104.39	96.99	15.22	8.11	1.65	202.03	6.66	6.89
VII	122.01	101.62	16.11	7.78	2.32	297.46	7.13	7.41
Grand mean	155.29	127.13	24.20	10.97	2.44	215.85	7.06	7.23

Table 62. Mean performance of each cluster of parents and hybrids in Ganesh × Coimbatore

Character/ cluster	Plant height (cm)	Plant spread East-West (cm)	Number of hermaphrodite flowers per plant	Number of fruits per plant	Fruit yield (kg)	Single fruit weight (g)	Fruit size	
							Length (cm)	Breadth (cm)
I	168.49	127.77	27.87	13.80	3.15	229.19	7.02	7.19
II	240.99	210.49	38.57	13.86	4.31	311.49	7.87	7.77
III	219.75	161.85	37.00	18.00	2.87	163.30	6.20	6.60
IV	128.27	96.87	19.67	9.00	2.64	294.07	7.47	7.67
V	116.14	96.87	26.47	15.20	2.79	184.33	6.43	6.67
VI	216.20	85.27	37.00	16.00	4.38	274.00	7.60	7.80
VII	243.90	107.10	41.00	12.00	4.63	386.00	9.15	9.00
Grand mean	191.32	215.70	32.51	13.98	3.54	263.20	7.53	7.53

Table 63. Mean performance of each cluster of parents and hybrids in Ganesh × Bedana

Character/ cluster	Plant height (cm)	Plant spread East-West (cm)	Number of hermaphrodite flowers per plant	Number of fruits per plant	Fruit yield (kg)	Single fruit weight (g)	Fruit size	
							Length (cm)	Breadth (cm)
I	235.77	177.88	40.73	24.18	3.99	165.96	6.46	6.66
II	191.77	109.54	17.71	10.29	2.10	203.05	6.65	6.87
III	241.63	211.64	39.11	13.44	4.38	328.04	8.16	8.04
IV	122.58	111.02	12.80	6.60	1.11	168.22	6.38	6.66
V	243.90	144.75	24.44	14.56	3.48	239.00	7.01	7.29
VI	83.40	61.20	9.00	4.00	0.38	96.00	5.60	6.00
VII	158.80	82.25	12.50	8.00	0.89	108.30	5.50	5.90
Grand mean	182.55	128.33	22.33	11.58	2.33	186.94	6.54	6.77

Table 64. Mean performance of each cluster of parents and hybrids in Ganesh × Yercaud-1

Character/ cluster	Plant height (cm)	Plant spread East-West (cm)	Number of hermaphrodite flowers per plant	Number of fruits per plant	Fruit yield (kg)	Single fruit weight (g)	Fruit size	
							Length (cm)	Breadth (cm)
I	278.90	167.64	26.40	15.40	3.05	198.34	6.75	6.90
II	109.39	86.59	17.24	9.00	1.70	186.55	6.58	6.78
III	149.47	112.72	27.79	14.50	3.20	221.36	7.01	7.21
IV	190.46	141.79	25.71	12.43	3.14	287.49	7.23	7.40
V	102.46	74.60	17.40	8.60	0.86	94.10	5.92	6.10
VI	191.10	152.88	31.50	15.50	3.43	221.85	7.07	7.15
VII	143.10	91.20	20.00	9.00	2.43	270.40	7.20	7.40
Grand mean	166.41	118.20	23.72	12.06	2.54	212.01	6.82	7.40

Table 65. Mean performance of each cluster of parents and hybrids in Jyothi × Kabul

Character/ cluster	Plant height (cm)	Plant spread East-West (cm)	Number of hermaphrodite flowers per plant	Number of fruits per plant	Fruit yield (kg)	Single fruit weight (g)	Fruit size	
							Length (cm)	Breadth (cm)
I	157.08	109.34	35.54	18.46	3.12	172.40	6.47	6.56
II	172.73	142.46	27.09	14.00	3.18	228.34	7.14	7.34
III	190.45	140.48	26.00	12.83	3.10	279.23	7.05	7.28
IV	106.70	103.02	16.25	10.00	1.87	186.05	6.74	6.95
V	101.30	105.45	13.00	10.50	1.37	128.60	6.50	6.70
VI	160.10	198.20	26.00	15.00	3.54	236.20	7.10	7.40
VII	190.50	149.60	24.00	10.00	3.37	337.00	8.30	8.10
Grand mean	154.12	135.51	23.98	12.97	2.79	223.97	7.04	7.19

Table 66. Mean performance of each cluster of parents and hybrids in Yercaud-1 × Kabul

Character/ cluster	Plant height (cm)	Plant spread East-West (cm)	Number of hermaphro-dite flowers per plant	Number of fruits per plant	Fruit yield (kg)	Single fruit weight (g)	Fruit size	
							Length (cm)	Breadth (cm)
I	151.88	131.77	19.15	11.08	2.04	182.41	6.63	6.82
II	278.90	167.64	26.40	15.40	3.05	198.34	6.75	6.90
III	130.10	118.90	21.00	10.00	1.65	164.60	5.60	5.90
IV	168.12	155.77	26.67	12.67	2.97	234.32	7.17	7.42
V	123.65	102.57	20.91	12.64	2.14	171.46	6.59	6.72
VI	134.85	102.17	21.50	12.25	2.72	220.70	7.03	7.22
VII	160.82	109.86	39.50	19.90	3.29	165.84	6.33	6.41
Grand mean	164.05	126.95	25.02	13.42	2.55	2.55	6.59	6.77

9

Genetic Diversity in Mango [*Magnifera indica* (L.)]

Manivannan, K. and C. Vasugi

The mango is one of the choicest fruit crops of tropical and sub-tropical regions of the world. Its popularity and importance can easily be realized by the fact that it is referred as 'King of fruits' in the tropical world. India ranks first, both in area (2.51 million hectares) and production (15.02 million tonnes) of mango and it contributes to 42.1 per cent of world's mango production. Presently, India is home to more than 1000 mango cultivars differing in taste, aroma, size, fibre content and pulp characteristics.

Though more than 1000 mango cultivars are reported, only 25-30 are of commercial importance. Most of the important cultivars are not amendable to hi-tech cultivation practices and do not meet the requirements of modern horticultural production systems like precocity in bearing, dwarfing, regularity in bearing with high yield, resistant to diseases, pests, physiological disorders and good keeping quality. Hence, breeding attempts to develop an ideal mango variety would be dwarf, regular bearer with medium size fruits (250-300g). Additionally, it should be highly tolerant to various fungal and bacterial diseases. A variety suitable for pickling is also gaining the attention of the breeders. An ideal pickling type should have high acidity, high fibre, good texture and characteristic raw mango flavour.

The first and foremost step in developing an ideotype is to explore the vast diversity and systematic characterization for utilization in the breeding programme.

Considering the importance of conserving the available gene pool, systematic exploration works are being carried out by National Institute like Indian Institutes of Horticultural Research (IIHR, Bangalore), National Bureau of Plant Genetic Resources (NBPGR, New Delhi) and others to collect and conserve the germplasm in Field Gene Banks. Recently, in addition to the traditional germplasm conservation methods, cryopreservation is also gaining importance especially in the conservation of nuclear gene (pollen parent) diversity for future breeding.

Utilizing the conserved germplasm in the breeding programme requires precise information on the genetic relationships within such germplasm. Information on the genetic distance among the germplasm accessions will also help to avoid the duplicates, widen the genetic base of the core collections and ultimately help in preserving the valuable diversity.

Considering the difficulties involved in the traditional divergence studies based on morphological characterization (Rajan *et al.,* 1999 and Dinesh and Vasugi, 2002), new methods based on studies at the DNA level are incorporated into fruit breeding programmes in order to accelerate and optimize genotypes fingerprinting and to study of genetic relationship among cultivars.

A range of DNA markers, *viz.,* AFLP, SSRs, ISSR, RFLP and RAPD have already been in use for exploring the diversity of the global mango gene pool. Among these, STMS (Sequence Tagged Microsatellite Sites) are of particular importance to study the variability, identification, germplasm conservation, domestication and movement of germplasm (Viruel *et al.,* 2005).

In divergence analysis, where several characters are simultaneously measured, multivariate methods like D^2 analysis, cluster analysis, principal component analysis help to identify the diverse genotypes and to identify the duplicate accessions and to understand the character association and contribution towards diversity.

Asynchronous flowering is one of the major hurdles in transferring some of the important traits like salt tolerance from *M. zeylanica* and odour from *M. odorata* and some early bearing varieties of *M. indica*. The recent advance in the field of cryopreservation of pollen will be helpful to overcome the asynchronous flowering habit. Since the ability of cryopresevd pollen depends on the species and duration of storage (Ganeshan and Rajasekharan, 2000a), the effect of cryopreserved pollen has to be assessed and compared with fresh pollen. Very few attemps have been made on this line in mango.

In the era where free and open exchange of germplasm is moving towards controlled and facilitated exchange, management of germplasm information is essential. Hence, integration of biotechnology and information technology is recommended for efficient management, conservation and utilization of germplam by various stake holders. In the light of IPR issue also, development of mango information system assumes greater significance.

The recent exploration mission carried out by the Indian Institute of Horticulture has added 35 unique pickling types from Western Ghats region to the field gene bank. These are called as Appemidi types. These Appemidi types are used for tender whole fruit pickles called 'midi' in local language. Characterization and assessment of diversity is essential to utilize these unique accessions in crop improvement. No work has been done on characterization and genetic relatedness of these unique types with the commercial varieties and polyembryonic varieties.

Fruit shape

The fruit shape is an important trait from the marketing point of view. The variety Alphonso has got attractive fruit shape thus got a good export potential. Variation in fruit shape has been recorded by several workers. Description of twenty dessert varieties of mango has been made and different fruit shapes *viz.*, ovate oblique (Kalapady, Neelum, Karpooramanga, Baneshan, Killichundan and Chausa), roundish oblique (Mulgoa, Moovandan and Panchasara Varikka) and roundish (Mundappa and Rumani) were recorded (Nandini and Elizabeth, 2002). In the evaluation of 17 mango accessions of Florida, six accessions were found most suitable based on scoring system for shape (Knight, 1993).

Fruit size

Growth and development causes irreversible increase in fruit size and increase in protoplasm, which may occur due to increase in cell division and cell size. The process of cell elongation and cell division provides the basis for fruit growth. The division and enlargement of cell is a complicated process involving synthesis of many organic compounds such as protein, cellulose and nucleic acid (Kumar, 1999). Fruit size is an important trait, which decides the marketability.

Fruits weighing about 300 to 400g are preferred in various domestic and international markets. Variation in fruit size has been recorded by various mango workers. Desai and Dhandar (2000) recorded that average fruit weight ranged from 203.0 g in the variety 'Bemcorado' to 422.5g in 'RC-MS-1', while Singh and Gangwar (1985) found that the average fruit weight ranged from 102.0 to 335.7g in nine mango varieties. According to Nandini and Elizabeth (2002) the fruit weight ranged from 142.5 to 534.0 g in 20 varieties. Zaied *et al.* (2007) evaluated eight local mango accessions *viz.*, Alphonso, Mabrouka, Hendi Meloky, Langra, Dabsha, El-Kobbaneia, Khad El Gamel and El-Madam and found that Langra variety had the biggest fruit followed by Dabsha and El-Kobbaneia, while El-Madam and Khade El-Gamel had the smallest fruit.

Pulp content

Pulp content is one of the primary traits to be considered while initiating a hybridization programme. A good mango variety should have high ratio

(3.3: 4.0) of edible to non edible matters (Singh *et al,* 1990). Desai and Dhandar (2000) recorded maximum pulp content of 83.21 per cent in RC-MS-1, followed by Salcatta Musserat (74.24 per cent). The pulp content of Goan varieties ranged from 70.4 to 79.4 per cent and was comparable with that of Alphonso which had 70.0 per cent (Nandini and Elizabeth, 2002), pulp content. A pulp recovery of 75.64 per cent has been recorded in Totapuri (Gowda and Huddar, 2004).

Pulp colour

Pulp colour is an important character from the consumer's point of view. The pulp colour of Goan varieties varied from yellow to deep orange. Attractive yellow orange coloured pulp similar to Alphonso was recorded in Furtado, Bemcorada and RC-MS-1 (Desai and Dhandar, 2000). Yellow colour was observed in Fernandin, whereas the variety Mankurad had yellowish orange or light orange. Degree of intensity of orange colour could be an indicator for higher carotenoids. The total carotenoids could improve the natural appearance of fruit products (Kalra *et al.,* 1994). The yellow orange colour pulp of Alphonso is due to higher carotenoids (13497 □g 100 g-^{1}), which provides more Vitamin A (Gowda and Huddar, 2004).

Skin colour

The preferred external appearance could get special advantage in fresh fruit market especially in export. The variety Salcette Mussarat had attractive fruit colour: ranging from yellow-orange to dark red blush (Desai and Dhandar, 2000). Besides Bardez Mussarat, RC-MS-1 and Fernandin too had attractive skin colour. These could be used as sources of colour in breeding. The duplicate gene in mango has been found to be responsible for its red skin colour (Panhwar, 2005). Achievements in mango breeding and genetics have been reviewed by Iyer and Dinesh (1996).

Total soluble solids (TSS)

The total soluble solids (TSS) are a measure of sweetness, which determines the fruit quality and consumer preference. Increased fruit quality with a concomitant increase in yield is a desirable trait in any study. The TSS content of fruits increases with the progress of ripening. The increase in TSS might be due to loss in moisture and hydrolysis of polysaccharides into simple sugars, conversion of organic acids into soluble sugars and enhanced solubilization of insoluble starch and pectin in the cell wall and middle lamellae (Gupta and Bramachari, 2004; Baghel *et al.,* 1989 and Sharma *et al.,* 1990). Maximum soluble solids of 21.3° B has been recorded in Mankurad variety of Goa (Desai and Dhandar, 2000). The TSS range of 13.75° B (Prior) to 19.65° B (Panchasara Varikka) has also been reported (Nandini and Elizebeth, 2002). For pickling, the genotype No.35 was found suitable with a TSS of 8.1°B (Chaudhari *et al.*, 2006). The TSS of Alphonso was reported as 18.0°B (Gowda and Huddar, 2004) and Langra as 13.4°B (Zaied *et al.*, 2007).

Titratable acidity

The titratable acidity is a measure of fruit quality, which plays a great part in the acid sugar balance and consequently influences the taste and flavour of the fruit (Lakshminarayana *et al.*, 1980). The titrable acidity ranged from 0.28 per cent (Mankurad) to 0.38 per cent (Barder Musseret) in Goan varieties (Desai and Dhandar, 2000) and 0.11 per cent in Suvarnarekha to 0.74 per cent in Karpooramanga (Nandini and Elizabeth, 2002). Variation in titratable acidity has also been reported by Gowda and Huddar (2004). The genotype No. 35 had high acidity (4.40 per cent) and found most suitable for pickling (Chaudhari *et al.*, 2006). The acidity ranged from 1.51 per cent in Alphonso to 1.89 per cent in Dabsha (Zaied *et al.*, 2007). The decrease in acidity during ripening is due to the metabolism of organic acids, which are the major respiratory substrates (Hoggag, 1995).

pH

The pH and acidity have an inverse relationship. The pH of mango pulp ranged from 3.4 to 4.65 in different cultivars (Nandini and Elizabeth, 2002). A pH range of 4.15 (Totapuri) to 4.86 (Amrapali) was recorded by Gowda and Huddar (2004).

Total sugars

The sugars in mango comprise of sucrose, glucose, fructose and maltose (Anon., 1994). The total sugars increase with ripening of mango fruits and this might be due to the high activity of glycogenic enzyme fructose 1, 6, diphosphatase during ripening (Kumar and Selvaraj, 1990). The accumulation of total sugars during the process of ripening is a consequence of starch hydrolysis. The activity of alpha amylase, beta amylase and starch phosphorylase leads to conversion of starch in to sugars (Chundawat and Rao, 1981). In Goa varieties, the total sugars were in the range of 12.23 per cent in Bardez Musserat to 16.32 per cent in Mankurad (Desai and Dhandar, 2000). The sugar content ranged from 12.31 per cent in Swarnarekha to 16.75 per cent in Chausa (Nandini and Elizabeth, 2002). The total sugars were high in Mallika (12.95 per cent) and Amrapali (15.41 per cent) which was attributed to higher TSS of more than 22 ° B (Gowda and Huddar, 2004).

Reducing sugar

The reducing sugar content of mango fruits increases during ripening due to conversion of starch into reducing sugar and loss of starch during ripening process (Sagar and Khurdiya, 1996). The reducing sugar content ranged between 2.0 and 7.5 per cent in different mango varieties (Bhatnagar and Subramanyam, 1973).

Vitamin C

Ripe mango fruits are moderate in Vitamin C content (Selvaraj and Pal, 1988). The ascorbic acid content exhibits a gradual decrease during ripening process. The fall in ascorbic acid content is attributed to its oxidation and enzymatic degradation during ripening. The ascorbic acid content ranged from 12.45 to 45.60 mg $100g^{-1}$ in different mango varieties (Nandini and Elizabeth, 2002). Genotypes with high vitamin C content (101.7 mg 100 g^{-1}) had been found most promising for pickling (Chaudhari *et at.,* 2006). The smaller fruits had higher amount of ascorbic acid content than larger ones (Palaniswamy *et al.,* 1974) and it decreased with storage at room temperature (Gofur *et al.,* 1994). The vitamin C content varied with cultivars (Sanyal *et al.,* 1991). It ranged from 2.76 mg 100 g^{-1} (Mabruka) to 26.24 mg 100 g^{-1} (Dabsha) in fruit pulp (Zaied *et al.,* 2007).

Flavour

Flavour is an important trait which decides the acceptability of the final product and is contributed by volatile compounds. It has been defined as a sensory phenomenon depending upon taste, odour or aroma, appearance, temperature, sensation of heat and cold and texture or mouth feel affecting the sense of touch (Manay and Shadaksharaswamy, 1995). The volatile compounds are responsible for odour and contribute to overall flavour of fresh and processed fruit (Brackmann *et al.,* 1993). More than 285 volatile compounds have been reported from various mango cultivars, including monoterpenes, sesquiterpenes, esters, aldehydes, ketones, alcohols, acids, aliphatic hydrocarbons and aromatics (Shibamoto and Tang, 1990 and Olle *et al.,* 1997). Aroma of mango fruit has been reportedly influenced by various factors like cultivars (Selvaraj and Pal, 1988; Shibamoto and Tang, 1990 and Olle *et al.,* 1997) and fruit maturity (MacLeod *et al.,* 1988 and Bartley and Schwede, 1987). Selvaraj and Pal (1988) were of the opinion that the variety Alphonso had very strong aroma, while Totapuri was with moderate aroma.

Gan *et al.* (1981) and more recently by Degani *et al.* (1990) and Schnell and Knight (1992) have demonstrated the feasibility of using isozymes as biochemical markers in mango. Isozyme variation among mango trees of the same putative cultivar was described by Gan *et al.* (1981), thereby indicating that somatic mutation may occur frequently in certain mango clones. Enzyme polymorphisms were used by Degani *et al.* (1990) to differentiate among mango cultivars and to identify parentage of certain modern mango cultivars. Schnell and Knight (1992) have been able to differentiate zygotic from nucellar seedlings on the basis of enzyme polymorphisms. Although isozyme markers are a relatively simple tool for genetic analysis and linkage studies, it is unlikely that a sufficient number of isozymes will be found to saturate the genome completely and uniformly (Tanksley, 1983).

Molecular characterization of Mango germplasm employing STMS markers

The names of many cultivars are thought to be synonyms of the same cultivar which could lead to confusion among the breeders and farmers. Hence, a critical method for cultivar identification needs to be developed. Molecular markers are used to measure diversity at DNA level and are not much influenced by environmental conditions like morphological markers.

Use of molecular markers in germplasm characterization

The discovery of molecular markers has lead to the detailed genetic analysis and approaches to improvement of crop plants. It is differentiated into Protein markers, which were discovered first and DNA based markers. Protein markers had a disadvantage that the proteins are affected by the influence of environment so that DNA markers are developed. The DNA markers are again categorized into Restriction based markers [Restriction Fragment Length Polymorphism (RFLP)] and PCR based markers [Random Amplified Polymorphic DNA (RAPD), Amplification Fragment Length Polymorphism (AFLP), Microsatellites, Inter-Simple Sequence Repeats (ISSR)]. The DNA based markers like RFLP and RAPD provide excellent tools to study the genetic diversity, eliminate duplicates in germplasrn, study relationships, gene tagging, genome mapping, PVR (Plant Varietal Rights) purposes, etc. The advantages of DNA markers are direct evaluation of genetic material; numerous markers can be obtained in a single population; assessment is not affected by environmental condition; powerful to pick up even minute differences and it requires little amount of plant tissue for analysis.

Molecular markers are practically unlimited in number, remain unaffected by environment and growth conditions and are simply inherited. The success of molecular markers in characterizing and analyzing genetic diversity in mango has been demonstrated by a number of studies. (Schnell and Knight, 1993; Adato, 1995; Schnell *et al.,* 1995; Lopez-Valenzuela *et al.,* 1997; Ravishankar *et al,* 2000; Hemanthkumar *et al.,* 2001). The RAPD technique was used by Schnell *et al.* (1995), Ravishankar *et al.* (2000) and Hemanthkumar *et al.* (2001) for cultivar identification and analysis of genetic relatedness in mango. Schnell *et al.* (1995), found the maternal halfsibs to cluster together indicating the usefulness of RAPD markers in studying the genetic relationships. Ravishankar *et al.* (2000) grouped mango cultivars into four categories based on RAPD analysis. Hemanthkumar *et al.* (2001) observed a moderate level of genetic diversity among south Indian cultivars by using RAPD technique. The RAPD technique developed by Williams *et al.* (1990) is based on the amplification of multiple segments of genomic DNA using arbitrary primers (small DNA sequences usually 10 nucleotide long). It is one of the most simplest techniques of molecular markers. According to Ferreira and Grattapalgia (1998) the development of this technique brought a real democratization in the area of molecular markers.

Use of STMS markers in characterization of germplasm

The DNA markers can either present a low reproducibility (RAPDs) or are time consuming and expensive (AFLPs, RFLPs). Consequently new methods based on studies at DNA level, which are easy and reliable for identification should be developed and must be incorporated into fruit breeding programmes in order to accelerate and optimize genotype fingerprinting and study genetic relationships among cultivars.

In the recent years, microsatellite markers are gaining importance for molecular characterization and genetic diversity studies (Fang and Roose, 1997; Moussaoui, 2005; Rafalski *et al.,* 1996 and Gupta *et al.,* 1996). Microsatellites are ubiquitous components of Eukaryotic genome (Delsney *et al.,* 1983 and Tautz and Renz, 1984). However, the evidence of their abundance in plant genome and their polymorphic character was reported only a few years ago (Condit and Hubbell, 1991; Lagercrantz *et al.,* 1993; and Morgante and Olivieri, 1993). It consists of a tandem repeats of mono, bi, tri or tetra nucleotides in the eukaryotic genomes. These molecular markers present a very high polymorphism based on the number of nucleotide motif repeats (Morgante and Olivieri, 1993). Besides polymorphism, they have a co-dominant inheritance, reproduceable and easy to interpret, amplified via polymerase chain reaction. SSRs have been isolated and characterized in large number of animal and plant species (Powell *et al.,* 1996 and Lefort *et al.,* 1998). Microsatellite marker primer pairs have been developed in peach (Sosinski *et al,* 2000; Testolin *et al.,* 2000; Aranzana *et al.,* 2002 and Dirlewanger *et al.,* 2002) and then used in genetic diversity analysis for sweet cherry (Dirlewanger *et al.,* 2002), peach

(Aranzana *et al.,* 2002), banana (Graphin *et al.,* 1998) and apricot (Romero *et al.,* 2003 and Zhebentyayeva *et al.,* 2003)

The main disadvantage of these markers is the high cost involved in designing the oligonucleotide primers specific to the flanking sequences of the microsatellite DNA regions. SSRs have been shown to detect more polymorphic alleles per locus than allozymes (Terauchi and Konuma, 1994) or RFLPs (Roder *et al.,* 1994) and an increasing number of laboratories are exploiting the potential of these markers for genome analysis and mapping in plants. SSR markers have already been developed and applied for many cereals. It has also been employed with tropical plants like avocado (Lavi *et al.,* 1994), citrus (Kijas *et al.,* 1995) kiwifruit (Weising *et al.,* 1996), banana (Kaemmer *et al.,* 1997; Lagoda *et al.,* 1998; Crouch *et al.,* 1999 and Creste *et al.,* 2004), almond (Shiren *et al.,* 2007), citrus (Shahsavar *et al.,* 2007) and grapes (Upadhyay *et al.,* 2007).

Use of STMS markers in characterization of mango

The potentials of microsatellite markers for identification of mango cultivars has been suggested by Honsho *et al.* (2005). Among 36 mango cultivars tested, 29 cultivars showed a unique pattern by six primer sets, whereas, seven cultivars could not be identified because of genotype similarities. The observed and expected heterozygosity ranged from 0 to 0.83 and from 0.29 to 0.73 respectively.

Schnell *et al.* (2005) developed 15 microsatellite loci, and evaluated 59 Florida cultivars of mango including four related species from the USDA germplasm collection. Two loci were monomorphic and 13 were polymorphic with two to seven alleles per locus. Four departed significantly from Hardy-Weinberg equilibrium and had significant heterozygote deficiency. Nine loci exhibited significant linkage disequilibrium. Cross species amplification was successful in 4 related species. These loci are being used to investigate patterns of genetic variation within *Mangifera indica* and between closely related species.

Duval *et al.* (2005) developed microsatellite markers in which 28 primer pairs produced polymorphic amplification products for a diversity sample including 15 mango cultivars and two accessions from the related species *M. laurina* and *M. applanata.* Nineteen SSR loci with clear scorable patterns were chosen to study diversity in the mango germplasm bank of Guadalupe. The number of alleles ranged from three to thirteen with observed levels of heterozygosity ranging from 0.059 to 0.857.

Use of other (RAPD) markers in mango characterization

As the works pertaining to STMS markers in characterization of mango is limited, use of other markers like RAPD on mango is reviewed here.

Karihaloo *et al.* (2003) carried out the RAPD analysis of 29 Indian mango cultivars comprising popular land races and some advanced cultivars. A UPGMA dendrogram showed the majority of cultivars from northern and eastern region of India clustering together and separate from southern and western cultivars. Analysis of molecular variants revealed that 94.7 per cent of the genetic diversity in mango existed within regions. However, differences among regions were significant. Northern and eastern regions formed one zone and western and southern regions formed other zone.

Souza and Lima (2004) studied forty mango genotypes of the Embrapa Meio-Norte mango collection using RAPD markers with 32 random primers. Based on UPGMA clustering, the genotypes were grouped into two distinct groups and were further subdivided into two groups. The Boot strap consistency test however, indicated that only three groups *viz.,* Mallika and Amrapalli, Van Dyke, Palmer and Keitt and Irwin and Glenn were really divergent.

Schnell *et al.* (1995) was the first to attempt the use of RAPD markers to study the phylogenetic relationship in mango. Their study supported some of the groupings based on anatomical characters, but did not support relationships among others based on the morphological classification. They also reported that a complete analysis using more molecular markers and including isozymes and RFLPs would provide additional data on interspecific relationships. In another study on identification of cultivars and validation of genetic relationships in *Mangifera indica* (L.), 25 accessions of mango were examined for RAPD genetic markers with 80 decamer random primers. Of the 80 primers screened, 33 did not amplify, 19 were monomorphic and 28 gave reproducible, polymorphic DNA amplification patterns. Eleven primers were selected from the 28 for the study. The number of bands generated was primer and genotype dependent and ranged from one to ten. No primer gave unique banding patterns for each of the 25 accessions; however, ten different combinations of two primer banding patterns produced unique fingerprints for each accession. A maternal half-sib (MHS) family was included among the 25 accessions to see if genetic relationships could be detected. RAPD data were used to generate simple matching coefficients, which were analysed phenetically and by means of principal co-ordinate analysis (PCA). The MHS clustered together in both the phenetic and the PCA while the randomly selected accessions were scattered with no apparent pattern (Schnell *et al.,* 1995).

Phylogenetic relationship

Schnell *et al.* (1995) estimated genetic relationships among *Mangifera* spp. based on RAPD markers. Phylogenetic relationships among nine species were studied using RAPDs. Analysis was conducted using average taxonomic distance, unweighed pair group method using arithmetic averages (UPGMA) and principal component analysis. Ten selected primers produced 109 usable bands. RAPD generated clusters did not always agree with the taxonomic classification based on phenotypic traits. When the two subsections of the genus were analysed separately, the classification more closely agreed with traditional taxonomic relationships.

Diversity analysis based on phenotypic markers

Diversity analysis is important for deciphering genetic relationship and for the efficient management of germplasm. Establishing the identity of crop variety using diversity study has assumed greater importance for protecting plant breeder's and farmer's rights.

Clustering based on Mahanalobis distance

Multivariate analysis by Mahalanobis's D^2-statistic is a unique and common tool for discriminating distant gene constellations. Using measures of genetic distance, the genotypes are grouped into different clusters, so that genotypes falling in one cluster are less divergent than those falling in different clusters. The data recorded on 19 quantitative characters were subjected to the D^2 statistic of Mahalanobis and average intra- and inter-cluster distances were calculated. Tocher's method (Rao, 1952) applied to the Mahalanobis distance matrix was used to determine cluster formation. Clustering of genotypes based on Tocher's method resulted in seven clusters (Table 67). Among the various traits, the titrable acidity explained the highest variation (38.98 %). It was followed by total sugars (15.39 %) and pH (11.19 %). Rajan *et al.* (2009) reported that fruit characters like pulp weight (34.03 %), peel weight (22.65 %) and TSS (10.22 %) contributed more towards the genetic divergence.

Though the accessions selected by Vasugi (2010) were broadly grouped into acidic types (20) in the first cluster and less acidic types (18) in the second cluster. The accessions *viz.*, Alphonso, Dashehari, Himsagar, Kana Appe 1 and Aruna Gowda Appe were distinctly separated from each other and formed mono clusters. Out of 43 accessions, the presence of 38 accessions in two clusters (20 in cluster I and 18 in cluster II) indicated a low genetic base of these accessions. Except Appemidi, Balekoppa appe and Chansi Appe all other members of the first cluster had less than 15° Brix and more acidity. In the second cluster, except Hitalahalli Appe and Malange, all members had more than 15° Brix and less acidity. Kana Appel, which showed the smallest fruit weight and the highest bisexual flower percentage, was grouped separately. Similarly, Dashehari, which had the highest total sugars; Aruna Gowda Appe, which exhibited the highest acidity; Himsagar, which possessed the highest stone weight and Alphonso, which differed from Dashehari in the percent of bisexual flowers, were grouped as mono clusters. Since, Mahanalobis distance analysis uses the covariance matrix for distance calculation, accessions with extreme values for important characters might have been separated from the rest as mono cluster.

Table 67. Composition of clusters based on D^2 analysis

Cluster	Name of Genotypes
I	Dannalli Appe, Halasage, Nandgar Appe, Holekoppada Appe, Sadamidi, Dodder Shida-dakke Appe, Huliappekai, Chanshi Appe, Mahabalagiri Appe, Adderi Jeerig Isagoor Appe, Jeerige, Kashimidi, Appemidi and Balekoppa Appe
II	Gaddalahalli Appe, Mani Bhatta Appe, Kovesara, Puttu, Kalwagudda, Kadikai, Muregeer, Hitalahalli Appe, Malange, Vattam, Tatamidi, Totapuri, Kangaramath
III	Kana Appe 1
IV	Dashehari
V	Aruna Gowda Appe
VI	Himsagar
VII	Alphonso

In highly heterozygous crops like mango, the low genetic base could be attributed to minor difference in the allelic pattern at different loci. Hence, minor differences between the varieties, like the acidity, can be explained. This probably is the basis for intervarietal differences. Even in the appemidi types, probably the present day cultivars would have a lineage in two different basic cultivars. Those varieties that have come from lesser acidic types have been grouped in one cluster and with high acidic types have been grouped in another cluster. Interestingly, the north Indian variety Dashehari, western variety Alphonso, and eastern variety Himsagar were identified as genetically distant from each other. This goes in agreement with the findings of Ravishankar *et al.* (2000) who stated that geographical diversity was closely associated with genetic diversity. On the contrary, Karibasappa *et al.* (1999) and Sulkeri *et al.* (1994) opined that the geographic diversity may not necessarily be related to genetic diversity.

It is possible to evaluate the differences between the studied traits using the means of the clusters formed in the cluster analysis (Table 68). The data in Table

68 presents the mean inter and intra cluster distances established by Tocher's optimization method. Intra cluster distance was always smaller than the inter cluster distance. The inter-cluster distance ranged from 56.11 to 136.57. The maximum inter-cluster distance (136.57) was observed between the clusters V (Aruna Gowda Appe) and VI (Himsagar), which revealed that these genotypes are more diverse. Cluster means of 19 different traits under study revealed that maximum fruit weight was observed for cluster VI (Himsagar), followed by cluster II and cluster VII (Alphonso). TSS was maximum in cluster IV (Himsagar) followed by III (Kana Appe-1) and VII (Alphonso) and acidity was maximum in cluster V (Aruna Gowda Appe) (Table 69).

Table 68. Average intra and inter cluster D^2 and D values for 43 mango accession

Cluster	I	II	III	IV	V	VI	VII
I	1626.11	6064.83	5848.58	12160.71	3641.04	14702.78	9201.99
	0.28)	(40.33)	(77.88)	(76.48)	(60.34)	(121.26)	(95.93)
II		1745.99	3210.02	3148.56	11513.29	4707.88	3382.47
		(41.79)	(56.66)	(56.11)	(107.30)	(68.61)	(58.16)
III			0.00	5503.86	14062.89	10178.39	3723.81
			(0.00)	(74.19)	(118.59)	(100.89)	(61.02)
IV				0.00	18291.48	2423.59	3762.96
				(0.00)	(135.25)	(49.23)	(61.34)
V						0.00	18650.55
						(0.00)	(136.57)

Clustering based on Euclidean distance

Since Mahanalobis D^2 analysis grouped many accessions which were morphologically diverse into a single cluster, a different distance measure (Euclidean) and clustering algorithm (UPGMA) was also attempted to study the genetic diversity. UPGMA analysis of 19 quantitative traits grouped the 43 accessions into two main clusters i.e., clusters A with 16 accessions and B with 27 accessions. It is interesting to note that those accessions that had a higher fruit weight than the overall mean (181.4g) were grouped in A and the rest were clustered in B. The high correlation between fruit weight and fruit quality traits observed in the present investigation also supports the classification. As a result of this significant correlation, all members of A had a high TSS. Cluster A was further divided into three sub-clusters, namely A_b, A_2 and A_3, and cluster B was sub-divided into B_b, B_2, B_3 and B_4. It was found that any two

Table 69. Cluster means of D^2 cluster and contribution of characters towards divergence.

Cluster/Character	I	II	III	IV	V	VI	VII	Number of times ranked first	Contribution towards % divergence
Fruit length (cm)	7.90	9.03	4.05	10.50	10.50	10.30	8.80	1	0.11
Fruit breadth (cm)	5.64	6.88	3.03	6.40	5.90	7.40	7.40	2	0.22
Fruit thickness (cm)	4.94	6.13	2.75	5.60	5.30	5.60	7.30	1	0.11
Fruit weight (g)	116.29	246.98	22.60	170.50	150.00	322.50	246.20	8	0.89
Pulp (%)	60.18	66.59	46.46	62.30	58.25	67.00	66.90	1	0.11
Titratable acidity (%)	4.31	0.61	1.31	0.13	7.48	0.19	0.39	353	38.98
pH	3.02	4.32	3.86	5.61	3.05	4.81	4.31	102	11.19
Vit C (mg 100 g^{-1})	75.06	40.02	60.00	18.29	89.90	12.80	23.80	63	6.98
Total sugar (%)	2.96	8.87	9.37	15.70	1.40	16.47	15.30	140	15.39
Non Reducing sugar (%)	1.74	5.26	5.68	9.23	0.86	9.87	8.37	1	0.11
Reducing sugar (%)	1.22	3.60	3.69	6.47	0.55	6.60	5.07	22	2.44
TSS °B	10.13	17.00	20.60	19.00	10.30	23.13	19.10	42	4.65
Leaf length (cm)	13.80	15.70	14.87	14.30	13.44	15.92	17.18	1	0.11
Leaf width (cm)	3.69	3.90	3.77	3.30	3.14	3.40	3.80	4	0.44
Petiole length (cm)	1.91	2.58	3.00	1.60	1.64	2.50	3.00	1	0.11
Inflorescence length (cm)	24.70	26.71	21.00	28.30	28.00	32.50	29.00	8	0.89
Stone length (cm)	6.38	7.13	3.50	7.63	9.20	10.47	6.03	62	6.87
Bisexual flower (%)	7.59	8.45	25.47	18.78	6.94	12.32	1.59	1	0.11
Stone weight (g)	26.91	31.29	7.00	24.70	47.00	55.00	22.43	95	10.41

genotypes falling under the same cluster showed the smaller genetic distance than those belonging to any other cluster. The total sugars (%), titrable acidity (%), pH, TSS (°B), stone length and vitamin C content decided the clustering of accessions into; A_1 A_2 and A_3 sub-clusters. Clustering of four accessions under A_1 was mainly decided by fruit weight (>130g), pulp per cent (>56), TSS (>15.0°B), total sugars (>7.0%) and moderate acidity (<1.0%) (Table 70).

The A_2 sub-cluster with three accessions *viz.*, Kangaramatha, Tatamidi and puttu were mainly decided by higher fruit weight (>220g), low acidity (<1.0%), high TSS (>17.0°B), medium sugars (>6.0%), yellow to yellow orange pulp with good eating quality. Clustering of six accessions *viz.*, Gaddalahalli Appe, Kalwagudda, Kovesara, Mani Bhatta Appe, Alphonso and Manipur under A_2 sub-cluster mainly shared the common traits like total sugars (>7.50%), low acidity (<0.6%), good TSS (>16.0°B), pulp per cent (>65 %) and orange to yellow orange pulp. Grouping of varieties like Dashehari belonging to North India, Himsagar of Eastern region and Vattam of South India in the same cluster indicated that there is no correlation between place of origin and clustering. Therefore, the selection of parental line for hybridization in mango should be based on genetic diversity rather than geographic diversity.

In the sub-cluster of A_3, all the three south Indian varieties *viz.*, namely totapuri, Banganapalli and Dantimamidi were grouped together, suggesting that they might have developed from common ancestors.

Clustering of 27 accessions under 'B' mainly consisted of pickling types. The first sub-clustering of B (was mainly decided based on fruit characters like good TSS (>24.0°B), sugars (>9.5 %), pH (>3.8), and moderate acidity (>0.5%). The second sub-clustering of B_2 with four accessions was mainly decided based on low sugars (<3.0 %), moderate TSS (<11.0°B) and high acidity (>3.0 %). The B_3 sub-cluster consisted of nine accessions that shared the common trait of moderate to high acidity (>0.6 %) with a pH of >2.5 and small-sized fruits. The accessions, Isagoor Appe, Jeerige and Thumbeebeedu shared all morphological traits like leaf, flower and fruit and are placed very closely in the same cluster, indicating the possibility of cultivar synonymy. The elimination of duplicate accessions helps in efficient management of field gene bank. Thus, fruit characters are observed to be the most important for explaining the genetic variability in mango. Thus, genetic diversity available in the gene pool could be exploited in the hybridization. The results obtained here also confirm the earlier report by Karibasappa *et al.* (1999) and Suman *et al.* (1985) that fruit characters were the major contributors for genetic diversity.

Table 70. Composition of cluster as per UPGMA clustering

Cluster	Accession	Cluster	Accession
A_1	Kadikai	B_3	Appemidi
	Dashehari		Balekoppa Appe
	Himsagar		Chansi Appe
	Vattam		Hittalahalli Appe
A_2	Kangaramatha		Isagoor Appe
	Tatamidi		Jeerige
	Puttu		Malange
	Gaddalahalli Appe		Shidadakke Appe
	Kalwagudda		Thumbeebeedu
	Kovesara	B_4	Adderi Jeerige
	Manibhatta Appe		Dannalli Appe
	Alphonso		Dodderi Jeerige
	Manipur		Gurumurthy Appe
A_3	Dantimamidi		Halasage
	Banganapalli		Holekoppada Appe
	Totapuri		Huliappe Kai
B_1	Kalkuni		Kalakai
	Kana Appe-1		Kashimidi
	Muregeer		Mahabalagiri Appe
B_2	Anantha Bhatta Appe		Nandgar Appe
i	Aruna Gowda Appe		
	Gorana Appe		
	Sadamidi		

The grouping of accessions in B_4 shared the place of origin as a common trait. Dannalli Appe, Halasage, Kashimidi, Mahabalagiri Appe, Nandgar Appe and Dannalli Appe which were from Sirsi; two accessions are from Sirsi, two accessions *viz.*, Dodderi Jeerige and Adderi Jeerige from Sagar; one accession *viz.* Holekoppada Appe from Siddapur and one accession Gurumurthy Appe from Kumta are grouped together. This indicates that probably the pickling (Appemidi) types existing within a geographical region are from a restricted gene pool and the present variability could have arisen due to seed propagation of heterozygous plants (Table 70).

The clustering pattern obtained in the present study also revealed that there was no relation between geographic and genetic diversity, as all the clusters consisting of more than one genotype included accessions from different ecogeographic regions. Murty and Arunachalam (1966) stated that genetic drift and selection in different environments could cause greater diversity than geographic distance. Sulkeri *et al.* (1994) have also indicated that geographic diversity cannot always be used as an index of genetic diversity. It is quite possible that many of the genotypes obtained from different geographical sources might have a common origin.

Principal Component Analysis (PCA)

In a multivariate analysis like clustering where a large number of characters are considered, it becomes very difficult to visualize the clustering pattern. Hence, Principal Component Analysis (PCA) was performed to visualize the association among accessions in more detail. Among the 19 possible principal components, the last 14 components were found to be equal (□^{2} test). Of the remaining five components, the first three were selected based on the criteria of minimum eigenvalue of one and cumulative variation of more than 70 %. The first three principal coordinates explained 70.89 % of the total variation. First, second and third principal coordinates explained 41.36, 18.10 and 11.43% of the total variation, respectively. In order to reduce the number of variables loading a principal component, varimax rotation was adopted. The rotated factor loading shown was used to interpret the rotated factor pattern. TSS, acidity, vitamin C, pH, total sugars, reducing sugars and non-reducing sugars showed significant loading (more than 0.7) on factor 1. Fruit length, breadth, thickness and weight, stone length and stone weight had significant loading on factor 2. Leaf length, leaf width and petiole length exhibited significant loading on factor 3. The rotated factors also explained more than 70 % of variation. TSS, acidity, vitamin C, pH, total sugars, reducing sugars and non-reducing sugars accounted for 30.5 per cent of the variability. Fruit length, breadth, thickness and weight, stone length and stone weight accounted for 24.9 per cent of variability (Table 71).

Table 71. Cluster means as per UPGMA clustering

Character / Cluster	Ai	A_2	A_3	Bi	B_2	B_3	B_4
Fruit length (cm)	9.38	9.38	1.40	5.44	10.63	7.83	7.07
Fruit breadth (cm)	6.97	7.43	8.38	4.11	7.53	5.64	4.95
Fruit thickness (cm)	5.71	6.75	7.54	3.95	5.53	5.37	4.40
Fruit weight (g)	229.93	276.28	457.23	64.33	161.13	134.03	89.13
Pulb (%)	65.32	71.17	68.70	51.84	60.79	64.99	57.24
Titrable acidity (%)	0.31	0.47	0.25	0.89	5.11	1.60	4.76
pH	5.03	4.36	4.55	4.10	2.99	3.31	2.98
Vitamin C (mg l00g^{-1})	24.54	41.33	21.77	58.31	82.32	58.40	78.27
Total sugar (%)	13.28	8.83	11.12	9.55	1.83	5.80	2.10
Non reducing sugar (%)	7.90	5.19	6.56	5.67	1.14	3.34	1.26
Reducing sugar (%)	5.37	3.44	4.49	3.87	0.69	2.46	0.84
TSS (°B)	17.43	21.15	19.13	24.52	9.45	14.48	8.94
Leaf length (cm)	16.43	16.47	13.90	16.43	12.86	14.43	13.46
Leaf width (cm)	3.93	4.05	3.33	4.01	3.18	3.94	3.57
Petiole length (cm)	2.58	2.88	1.80	2.94	1.69	2.12	1.82
Inflorescence length (cm)	34.21	25.80	25.62	21.23	28.00	26.96	22.78
Stone length (cm)	8.20	7.06	8.83	4.63	7.91	6.17	6.25
Bisexual flower (%)	11.56	6.12	3.94	21.01	5.70	8.31	7.21
Stone weight (g)	35.28	32.96	35.37	16.59	4Q.85	25.42	24.70

Leaf length, leaf width and petiole length accounted for 15.5 per cent of variability. Percentage contribution of variables towards divergence observed in D^2 analysis was in close agreement with the PCA.

Rotated factor scores of 43 accessions indicated that all the genotypes were clustered around the centre. The accessions Kana Appe-1, Muregeer and Himsagar were identified as outliers. Accessions with high TSS and sugars were found at the top (Jeerige, Puttu, Tatamidi, Manipur, Balekoppa Appe, Thumbeebeedu, Malange, Kalwagudda, Gaddalahalli Appe, Kovesera, Totapuri, Kalkuni, Dantimamidi, Mani Bhatta Appe, Kana Appe 1, Muregeer, Alphonso, Banganapalli, Vattam and Himsagar).

These less acidic accessions were further sub-divided into four subgroups, based on their scores on factor 2 and factor 3. Malange, Vattam, Mani Bhatta Appe, Thumbeebeedu and Jeerige were grouped in sub-cluster 1. Gaddalahalli, Alphonso, Kana Appel, Kovesara, Balekoppa Appe, Muregeer, Manipur and Kalkuni formed sub-cluster 2. Banganapalli, Dashehari, Totapuri, Himsagar and Dantimamidi were found in the third sub-cluster. Puttu, Kalwaguda and Tatamidi were grouped in the fourth sub-cluster.

Table 72. Contribution of Principal Components towards divergence

Principal Component	Eigen value	Total Variance (%)	Cumulative Eigen value	Cumulative (%)
1	7.85	41.36	7.85	41.36
2	3.43	18.10	11.29	59.46
3	2.17	11.42	13.46	70.89

Composition of Principal Component cluster

Character	Factor 1	Factor 2	Factor 3
Fruit length (cm)	0.095	0.901	-0.016
Fruit breadth (cm)	0.306	0.844	0.040
Fruit thickness (cm)	0.329	0.806	0.278
Fruit weight (g)	0.445	0.781	0.162
Pulp (%)	0.221	0.555	0.480
Titratable acidity (%)	-0.774	-0.124	-0.310
pH	0.841	0.246	0.228
Vitamin C (mg 100g^{-1})	-0.619	-0.122	0.036
Total sugar (%)	0.962	0.136	0.089
Non reducing sugar (%)	0.963	0.134	0.080
Reducing sugar (%)	0.963	0.135	0.094
TSS (°B)	0.782	0.052	0.353
Leaf length (cm)	0.199	0.131	0.893
Leaf width (cm)	0.008	-0.103	0.851
Petiole length (cm)	0.269	-0.002	0.882
Inflorescence length (cm)	0.220	0.269	0.034
Stone length (cm)	0.083	0.762	-0.075
Bisexual flower (%)	0.288	-0.471	0.036
Stone weight (g)	0.016	0.771	-0.068

Accessions with high acidity were grouped at the bottom (Arunagowda Appe, Halasage, Holekoppada Appe, Nandgar Appe, Dannalli Appe, Anantha Bhatta Appe, Gorana Appe, Kalakai, Kashimidi, Dodderi Jeerige, Sadamidi, Shidadakke Appe, Mahabalagiri Appe, Gurumurthy Appe, Appemidi, Isagoor Appe, Added Jeerige, Huliappekai, Chansi Appe, Kadikai, Kangaramatha and Hittalahalli Appe).

Gurumurthy Appe, Mahabalagiri Appe, Dodderi Jeerige, Chansi Appe and Isagoor Appe formed the first sub-group. Nandgar Appe, Kalakai, Kashimidi, and Shidaddake Appe formed the second sub-group. Sadamidi, Holekoppada Appe, Added Jeerige, Huliappekai, Anantha Bhatta Appe, Arunagowda Appe, Halasage, Gorana Appe and Hittalahalli Appe constituted the third sub-cluster. Dannalli Appe, Kangaramatha, Appemidi and Kadikai comprised the fourth sub-cluster.

Commercial varieties of different regions behave differently when grown in other agroclimatic zones. The variety Amrapalli which remains dwarf under North Indian conditions does not remain so under South Indian conditions. Dashehari, Langra and Chausa rarely flower under South Indian conditions. Yadav and Singh (1985) opined that South and North Indian varieties belong to two different ecotypes of *M. indica* based on physiology of flowering. In such situations diversity analysis based on morphological characters are likely to show inconsistencies. It should be noted that in the present investigation biochemical parameters like acidity, TSS, total sugars, reducing and nonreducing sugars (principal component 1) had significantly contributed towards divergence than the fruit characters (principal component 2). Hence, concentrating on few characters as described before would result in consistent clustering pattern.

Compared with D^2 analysis, Euclidean distance-based clustering had indicated logical clustering of similar genotypes which shared some common traits. Pie better clustering observed in Euclidean clustering could be attributed to the distance measure which uses the standardized data and correlation matrix rather than covariance matrix used in Mahanalobis distance estimation. Further, it might be to unweighted pairwise group method based on arithmetic mean. Thus, for a univariate data set involving variables measured in different units, Euclidean stance could be recommended for studying the diversity.

Molecular characterization of mango germplasm employing STMS markers

The traditional method of morphological characterization is a common step in plant breeding although it has some drawbacks, as it is descriptive, error-prone, and affected by the environmental factors. Therefore, several molecular tools are being adopted for plant variety characterization and identification because these techniques are reliable, unambiguous in nature and easy to adopt (Karp *et al.*, 1997).

The discovery of molecular markers in recent years has greatly enhanced the scope for detailed genetic analysis and approaches for the improvement of crop plants. Molecular markers are widely used for the estimation of genetic diversity and are reported to be more precise than phenotypic markers (Gupta and Varshney, 2000). Among the markers, the microsatellite DNA markers have been the most widely used as they have co-dominant inheritance, reproducibility and are easy to interpret (Powell *et al,* 1996 and Lefort *et al.,* 1998). In mango, microsatellites have been successfully used for genetic diversity analysis by Schnell *et al.* (2005), Viruel *et al.* (2005) and Lopez *et al.* (2009).

In the present investigation, the sequence-tagged microsatellite markers were successfully used to estimate the genetic distances between 43 mango accessions. A total of 14 STMS markers developed at the Institute were employed in the analysis of molecular characterization variation of the mango accessions.

The analysis of 14 STMS markers resulted in the detection of a total of 211 alleles. The allele size ranged from 108 to 153 bp in Mi-IIHR 23 to 230 to 269 bp in Mi-IIHR 17. The expected heterozygosity ranged from 0.643 in Mi-IIHR-18 to 0.931 in MiIIHR-17. The observed heterozygosity (0.881) and number of alleles per locus (23) was more in Mi IIHR-17 which explained its high polymorphism. The ability of STMS markers in distinguishing mango cultivars was previously reported by several workers (Viruel *et al.,* 2005 and Lopez *et al.,* 2009). The allelic data obtained f\rom gene scan analyser was first converted into a binary format using band size to binary software. The binary data then used to calculate the similarity coefficient (dice coefficient) and heirarchial clustering was performed based on UPGMA. The correlation between clustering and original data was tested using 1000 replicate bootstrap analysis. According to Felsenstein (1993), bootstrap values higher than 50 per cent may be considered as statistically significant.

The UPGMA clustering technique separated the accessions into less acidic and acidic groups. Accessions with moderate acidity, medium to high TSS and with moderate level of sugars formed a separate cluster (A_2). Malange and Appemidi were found to have been the part of A_2, which showed similarity in fruit weight (100 to 111 g), dense lenticels with strong turpentine flavour. A close relationship was also noticed between the accessions Balekoppa Appe and Shidadakke Appe, which share the common trait of oblong fruit shape with smooth skin. The results obtained in the present study indicated a relatively close relation between the accessions Sadanudi and Huliappekai. Morphologically, also these two shared the common quality traits of low sugar (<2.5%), TSS (<10.0°B) and pH (>2.5).

The genomic similarity as evidenced by the close relationship between the accessions *viz.,* Gaddalahalli Appe, Mani Bhatta Appe and Muregeer to form a

Table 73. Allelic composition of 43 mango accessions using STMS markers (Mi IIHR 19 – Mi IIHR 32)

S-No	Cultivars/ Markers	MiIIHR 19a	MiI-IHR 19b	MiI-IHR 24a	MiI-IHR 24b	MiI-IHR 18a	MiI-IHR 18b	Minim 26a	MillHR 26b	MiI-IHR 15a	MillHR 15b	MillHR 31a	MillHR 31b	MillHR 32a	MillHR 32b
1	ADDERIJEERIGE	185	203	238	254	168	168	155	157	153	153	208	217	183	183
2	ANANTHA BHATTA APPE	191	199	238	238			155	165	136	136	208	217	183	183
3	APPEM1DI	185	191	239	244	168	168	150	155					183	183
4	ARUNA GOWDA APPE	185	201	238	248	168	168	155	155	137	137	208	217	18C	182
5	BALEKOPPA APPE	185	191	246	254	168	168	143	155	152	152	208	211	183	191
6	CHANSHI APPE	191	195	238	238	168	181	137	137	152	156	208	217	195	195
7	DANNALLI APPE	185	202	238	248	163	163	137	155	148	148			181	18 i
S	DANTIMAMIDI	187	195	244	248	168	168	137	143	131	137	220	229	183	195
9	DODDERI JEERIGE	189	198	239	239	168	168	156	156	162	168	209	218	183	186
10	GADDALAHALLI APPE	191	195	244	244	168	168	143	161	128	156	211	220	183	192
11	GORANA APPE	185	197	238	238	161	161	153	153	151	151	217	223	183	183
12	GURUMURTHY APPE	185	201	238	254	163	163	155	155	151	151	207	207	183	183
13	HALASAGE	197	209	238	238	158	158	139	155	152	155	223	223	181	183

14	HITT ALAH ALLI APPE	191	191	239	248	161	1681	155	155	136	136	208	211	ISI	191
15	HOLEKOPPADA APPE	197	199	253	257	161	161	155	162	148	151			181	181
16	HULIAPPEKAI	199	199	238	238	168	193	135	155	112	112	208	208	181	181
17	ISAGOOR APPE	190	197	238	244	168	168	157	163	136	139	220	223	183	183
18	JEERIGE	191	197	239	245	163	163	157	162	136	139	220	220	183	183
19	KADIKAI	185	187	244	248	168	168	145	157	151	151	220	223	181	187
20	KALAKAI	189	199	238	238	168	168	137	157	136	159	208	217	183	187
21	KALKUNI	197	197	239	239	155	155	155	155	111	136	217	217	181	181
22	KALWAGUDDA	187	195	244	244	168	168	137	151	151	155	220	220	192	192
23	KAN A APPE-1	185	205	258	258	161	161	135	135	148	152	211	220	187	189
24	KANGA RAMATHA	185	195	244	244	168	168	137	163	155	155	223	229	180	191
25	KASHIMIDI	191	197	239	258	168	168	155	155	142	142	208	214	177	183
26	KOVESARA	185	185	248	248	168	168	137	161	136	152	220	220	177	183
27	MAHABALAGIRI APPE	197	209	238	254	168	168	151	155	111	136	208	217	182	182
28	MALANGE	185	191	244	248	168	168	151	155	137	137	211	211	183	183
29	MANI BHATTA APPE	185	195	245	245	161	165	143	155	126	126	220	220	183	192
30	MUREGEER	185	185	238	244	168	168	143	155	127	152	208	223	183	187
31	NANDGAR APPE	185	187	238	254	168	183	155	155	137	156	217	217	181	183

32	SADAMIDI	197	201	238	254	163	163	155	155	136	136	208	208	181	183
33	SHIDADAKKE APPE	185	187	244	254	166	168	143	155	137	152	217	220	183	191
34	TATAMIDI	185	185	249	249	168	168	136	136	156	156	221	260	182	188
35	THUMBEEBEE-DU	191	197	239	244	168	168	157	163	137	140	220	220	183	183
36	ALPHONSO	185	185	244	248	167	167	144	162	151	151	220	229	182	186
37	BANGANAPALLI	195	204	245	245	168	168	136	156	152	152	217	230	183	192
38	DASHEHARI	194	200	244	260	154	154	136	144	154	154	220	220	192	192
39	H1MSAGAR	187	191	244	248	172	174	146	146	153	156	223	223	183	195
40	TOTAPURI	187	187	244	260	167	167	144	144	151	151	220	220	182	196
41	PUTTU	195	206	244	244	155	155	145	147	127	127	220	229	192	192
42	VATTAM	191	195	238’	245	168	168	131	157	137	153	217	220	181	183
43	MANIPUR	195	195	244	244	168	168	131	136	151	151	220	220	181	183

Table 74. Allelic composition of 43 mango accessions using STMS markers (Mi IIHR 23- Mi IIHR 34)

S No	Cultivars/ Markers	MillHR 23a	MillHR 23b	Min HR 30a	MillHR 30b	MillHR 36a	MinHR 36b	MinHR 17a	MillHR 17b	MillHR 13a	MinHR 13b	MillHR 12a	MillHR 12b	MillHR 34a	MillHR 34b
1	ADDER1JEER1GE	118	128	190	209	239	241	246	246	183	183	175	175	223	235
2	ANANTHA BHAT-TA APPE			195	198	239	249	244	263	182	182	175	175	222	234
3	APPEMIDI	149	149	198	213	235	240	239	243	169	188	175	177	222	235
4	ARUNA GOWDA APPE	139	139	195	197	239	239	236	252	183	183	173	173	225	235
5	BALEKOPPA APPE	139	147	199	207	227	239	237	250	181	188	177	175	222	234
6	CHANSHI APPE	147	147	199	205	221	221	238	238	183	183	169	175	222	237
7	DANNALLI APPE			196	196	220	239	244	251	181	181	173	175	228	228
8	DANTIMAMID1	147	147	190	198	221	226	243	245	181	188	169	175	237	244
9	DODDERI JEERIGE	118	128	194	199	221	241	247	251	183	183	154	174	223	223
10	GADDALAHALL1 APPE	141	153	199	201	227	245	236	243	181	188	169	175	228	228
11	GORANA APPE ‘			196	196	237	237	244	269	181	181	154	173	222	234
12	GURUMURTHY APPE	141	141	195	199	239	239	244	252	182	182	175	175	228	228
13	HALASAGE	118	129	197	199	223	239	245	263	182	188	177	175	222	228

14	HITTALAHALLI APPE	153	153	194	207	239	239	239	252	182	189	175	175	228	228
15	HOLEKOPPADA APPE	138	138	198	198	239	247	244	251	181	181	154	175	222	225
16	HULIAPPEKAI	137	137	197	199	219	239	252	264	183	183	169	175	222	222
17	ISAGOOR APPE	139	139	199	209	241	247	236	245	182	188	154	169	222	228
18	JEERIGE	128	128	199	209	241	248	237	245	183	190	154	169	222	228
19	KADIKAI	147	147	195	199	229	241	243	264	189	189	169	175	222	238
20	KALAKAI	131	131	195	199	221	241	248	252	182	188	154	173	222	222
21	KALKUNI	135	135	199	199	239	239	245	252	181	181	154	173	228	235
22	KALWAGUDDA	143	128	199	207	221	235	236	240	188	188	169	169	238	238
23	KAN A APPE-1	128	145	194	194	210	218	230	236	183	183	154	175	222	222
24	KANGARA-MATHA	147	147	197	199	221	247	252	252	182	188	169	169	228	237
25	KASHIMIDI	118	128	198	198	239	239	245	252	183	183	175	175	222	222
26	KOVESARA	147	153	197	201	221	245	237	245	182	182	169	175	225	228
27	MAHABALAGIRI APPE	143	147	194	199	235	239	252	258	183	183	154	154	234	244
28	MALANGE	118	128	195	205	235	239	243	264	182	188	169	175	222	235
29	MANIBHATTA APPE			197	197	227	239	237	243	188	188	169	169	222	228
30	MUREGEER	134	153	197	197	227	239	230	243	183	183	169	175	228	238
31	NANDGAR APPE	118	128	195	199	239	239	245	252	182	188	154	169	228	238
32	SADAMIDI	151	134	194	197	239	239	252	264	183	183	169	175	222	228

33	SHIDADAKKE APPE	147	147	207	207	227	239	243	252	183	183	175	177	225	228
34	TATAMD1	134	146	197	197	221	221	237	242	183	189	169	169	238	238
35	THUMBEEBEE-DU	139	139	199	209	241	247	236	245	181	188	154	169	222	228
36	ALPHONSO	128	140	196	196	229	247	236	243	181	187	169	175	238	238
37	BANGANAPALLI			195	195	221	241			182	189	173	173	225	238
38	DASHEHARI	128	146	198	198	220	228	266	266	187	187	169	169	234	244
39	HIMSAGAR	141	147	199	205	231	231	241	245	188	188	167	175	228	234
40	TOTAPURI	148	148	190	207	228	228	266	266	181	182	169	175	222	228
41	PUTTU	147	149	194	198	231	231	240	245	181	181	169	175	234	234
42	VA1TAM	108	128	195	201	214	241	239	265	182	188	169	175	228	238
43	MANIPUR			199	201	215	221	239	251	182	188	169	175	228	228

sub cluster A_2 Yellow orange pulp with slightly juicy nature of pulp combined with moderate sugar (>6.8%) and medium TSS (>13.5°B) were noticed as the common traits possessed by these accessions. The accession Hittalahalli Appe also had attractive yellow orange pulp and was found to have clustered together in A_2 sub-cluster. As the accessions evaluated in the present study are unique indigenous types, there are no previous reports to justify the present result. However, using molecular markers, Karihaloo *et al.* (2003), Souza and Lima (2004) and Schnell and Knight (1993) studied the genetic relationship in mango and found that there was considerable diversity in mango gene pool.

The accessions Thumeebeedu and Isagoor Appe were placed very closely indicating a high similarity between these two accessions. Morphologically also they shared all the traits and look alike, which has been proved at the molecular level clearly. These two accessions also had a relatively close association with Jeerige where changes in one or two loci have brought out minor changes in morphological trait, which has been explained in the present study. Similarly, Chansi Appe grouping with Himsagar showed the common trait of strong turpentine flavour, indicating the effect of STMS markers in qualitative traits. Thus, it is clear from the study that both qualitative and quantitative traits could be mapped using STMS markers (Tables 73, 74, and 75).

Clustering of accessions in "B" is mainly based on the qualitative traits like high TSS, low titrable acidity, low pH and medium to high sugar, making these accessions suitable for table purposes. The genetic identity of polyembryonic cultivars has been justified from the present study. The polyembryonic types *viz.,* Vattam, Puttu and Manipur were grouped in cluster "B" along with other mono embryonic cultivars. It has been noticed in the present study that there is a relatively close relation between the accessions Manipur and Vattam. Arnon *et al.* (1998), also suggested that both monoembryonic and polyembryonic genotypes are inter-crossable.

Interestingly molecular characterization also revealed no correlation between geographical origin and clustering pattern. Clustering of genotypes belonging to different geographic region suggests that they might have evolved from the existing mango gene pool from which they were selected by local people to domesticate them in different areas for cultivation. The findings of the present study is in agreement with the report published by Pandit *et al,* (2007) who also observed no geographic separation of North and South Indian mangoes. Bajpai *et al.* (2008) also reported that cultivar pool from north and south region of India is considerably homogeneous.

Table 75. Locus name, length of the amplified fragments and variability parameters in 43 mango cultivars using 14 STMS markers

SI. No	Locus		Exp.Size (bp)	No. of alleles	Allele size range (bp)	Expected heterozygosity	Observed heterozygosity	Polymorphic information content
1	FMID010	Mi-IIHR12	178	7	154-177	0.734	0.698	0.680
2	HMID074	Mi-IIHR13	190	7	169-190	0.798	0.442	0.757
3	HMID04	Mi-IIHR15	153	21	111-168	0.920	0.476	0.902
4	HMID034	Mi-IIHR17	244	23	230-269	0.931	0.881	0.915
5	HMID016	MHIHR18	168	14	154-193	0.643	0.167	0.617
6	FMI223	Mi-IIHR19	187	18	185-209	0.874	0.791	0.850
7	FMID036	Mi-IIHR23	144	20	108-153	0.916	0.567	0.897
8	HMID019	Mi-IIHR24	246	12	238-260	0.840	0.581	0.811
9	FMI250	Mi-IIHR26	166	20	131-165	0.877	0.674	0.858
10	FMI2007	Mi-IIHR30	199	12	190-213	0.887	0.698	0.864
11	HMI2041	Mi-IIHR31	229	13	207-260	0.829	0.600	0.798
12	FMI2025	Mi-IIHR32	188	13	177-196	0.786	0.595	0.758
13	HMI2046	Mi-IIHR34	237	9	222-244	0.818	0.674	0.784
14	HMI2038	Mi-IIHR36	246	22	210-249	0.873	0.689	0.852

It is clear from the results that STMS analysis is efficient to prove the genetic distances between mango accessions at the genomic level. This is well justified from a cluster where all the acidic types are clustered together. The ability of STMS markers in distinguishing different species of Mango was previously reported by Duval *et al.* (2005, 2006), Honsho *et al.* (2005), Schnell *et al.* (2005), Viruel *et al.* (2005) and Lopez *et al.* (2009).

Comparison of morphological and molecular markers in characterization vis-a-vis cultivar synonymy

The genetic distances between 43 mango accessions were estimated using both morphological and molecular markers. Both these methods were able to identify the acidic pickling types from the commercial types and poly embryonic types. The genetic identity of the accessions namely Dantimamidi, Puttu, Alphonso, Dashehari, Totapuri, Kalwagudda, Kangaramatha, Kadikai, Vattam, Manipur and Kovesara has been revealed by way of clustering into a single group in both methods.

As in case of morphological clustering, the accessions *viz.,* Isagoor Appe, Thumbeebeedu and Jeerige were grouped under a single cluster in the molecular dendrogram indicating their genetic identity. A casual observation of leaf, flower, fruit and stone characters indicated that Isagoor Appe, Thumbeebeedu and Jeerige may be duplicate accessions. However in the D^2 analysis they were grouped in different clusters. In the dendrogram obtained from Euclidean coefficient (morphological data) and dice coefficient (molecular data) both Isagoor Appe and Thumbeebeedu were clustered together. The differential clustering observed under D^2 analysis might have resulted due to heterogenous units used in the multivariate analysis. Vipin *et al.* (2009) also stated that clustering based on Euclidean coefficient which neutralizes the unit bias was more precise and reproducible. In addition, the molecular data, which is free from the environmental influence is highly reliable and accurate (Gupta and Varshney, 2000). In morphological analysis, the accessions Appemidi and Malange were grouped in B_2 sub cluster, which was proved by molecular markers. Similarly, the accessions Balekoppa Appe and Shidadakke Appe by the way of their acidic nature clustered in a single group in morpho dendrogram, was confirmed by molecular markers.

The accession Kashimidi was closely related to Hittalahalli Appe, which in turn clustered with Muregeer. Even though the accession Kashimidi was morphologically different in terms of size and shape but proved to be genetically similar in the present study. Thus, the present study clearly proves that the morphological characterization can be complemented by the molecular characterization. Similar results were reported by Weerarante *et al.* (2005). Higher efficiency of STMS markers in molecular characterization over morphological markers is well demonstrated (Duval *et al.,* 2005, 2006; Honsho *et al.,* 2005; Schnell *et al.,*2005; Viruel *et al.,*2005 and Lopez *et al.,* 2009) in mango.

Considering the above points it is concluded that Thumbeebeedu and Isagoor Appe and Jeerige are the one and the same.

Utility of phenotypic traits and molecular markers in diversity analysis

The goodness-of-fit of the UPGMA dendrograms generated with morphological and microsatlellite data were tested by Mantel test. Moderate support for clustering patterns was observed for two markers type (CCS=0.865, for morphological and CCS=0.71, for molecular data).The correlation coefficient and significance of the conelation matrices based on morphological and molecular data tested by the Mantel test showed that only moderate correlation existed between phenotypic distance and genetic distance ($r = 0.32$). Therefore, the present study suggested that phenotypic traits and molecular markers should be complemented with each other in order to achieve a reliable evaluation and robust characterization of the species diversity.

Possible gene donors for important traits.

Though the 35 accessions were catalogued as pickling types, perusal of the data on various morphological, biochemical characters and secondary descriptors like overall eating quality, maturity period and fruit attractiveness showed some promising accessions that could be considered for fresh fruit market. Apart from the existing commercial varieties included in the study, the accessions Dantimamidi and Kovesara were identified on par with the commercial varieties.

Further, the accessions Gaddalahalli Appe, Hittalahalli Appe and Kalkuni which had attractive orange skin could be used as gene donor for incorporating skin colour. Similarly, the accessions Hittalahalli Appe, Kangaramatha and Kovesara which posessed orange colour pulp could be exploited as gene donor for incorporating pulp colour.

10

Genetic Diversity in Jackfruit [*Artocarpus heterophyllus* (Lamk.)]

Manivannan, K. and R. Rajesh Kannan

Jack bearing the largest edible fruit in the world, is a tropical evergreen tree belonging to the family Moraceae. Previously, it was known as *Artocarpus integrifolia.* It is an indigenous fruit crop, probably originated in Western Ghats of India (Santapau, 1966) and is widely cultivated in South Asia, East Indies and other warm areas of both the hemispheres. Jack tree is commonly grown in Burma, Malaysia and Brazil. It is hardly recognized as a commercial fruit crop in India even though it is widely distributed in Southern states *viz*,. Kerala, Tamil Nadu, Karnataka and Andra Pradesh. It is also cultivated in other states like Assam, Bihar, Orissa, Maharashtra and West Bengal. It is an invariable component of Kerala's homesteads and also grown as a shade crop in coffee, arecanut and cardamom plantations. It is also grown as a standard crop in pepper plantations. The crop has got a rare distinction of being used as the staple food of Kerala tribals and also in some African countries like Uganda.

Owing to its numerous culinary uses and its availability in plenty during the heavy monsoon rains, jackfruit has earned the well deserved name "Poor man's food" (Sammadar, 1990).

Jack flakes are rich source of carbohydrate (16 to 20%), total soluble solids (29° brix), carotene (500 to 580 IU) and pectin (1.5 to 6%). The flakes contain moisture (77.2%), protein (1.9%), fat (0.1%), fibre (1.1%), mineral matters (0.8%), calcium (0.02%), phosphorus (0.03%), Fe (0.5 mg) and vitamin C (10 mg 100 g^{-1}) (Dutta, 1957).

The tree has good timber value as it is rarely attacked by white ants (Krishnankutty, 1998). The tree is extensively used in villages for furniture and building materials. The leaves provide an excellent feed for goats and stitched leaves are used as plates and containers. An yellow die is obtained by boiling matured woods or saw dust for dying garments especially for priests.

The skin of fruits, bark of tree and leaves yield white latex which is collected and used for caulking boats and as stickers (Dutta, 1957).

The flakes are used in both unripe and ripe stages. Apart from the table purpose, the ripe flakes can also be used for making canned products such as nectar, preserves jam, jelly, squash, fruit bar and candy. Unripe green jack flakes are used for making chips and papad. In Kerala, tender fruits of about 60 days old are used as vegetable. Seeds are rich source of starch and also form a popular ingredient in many culinary preparations. Roasted seeds, taste like chestnuts and boiled seeds are made into delicious sweets in Indian homes.

Recently, it has been reported that jackfruit could be very useful in the treatment of the dreaded diseases of human being such as AIDS. An extract of jack fruit called 'Jacaline' was seen to have inhibited the growth of HIV infection *'in vitro'*. Jacaline is inactive on lymphocytes which are already infected but has proved its might by protecting the healthy ones. Hot water extracts of leaves improve the glucose tolerance level of diabetic persons. Lectin, an important class of natural protein, plays a significant role in cancer treatment since it has been in the prediction of transformation of normality to cancer cells. In Ayurveda, it has been used for curing inflammation, constipations, skin disease and wound healing.

Jack is a highly heterozygous, cross pollinated crop as such seedlings exhibit a wide range of variations which aids in the selection of superior desirable genotype. Due to cross pollination and predomination of seed propagation over a large period of time, it shows innumerable types or forms consisting the fruit characteristics. The types differ widely among themselves in spine on the rind, bearing, size and quality, period of flowering and maturity. Wide variations of jack fruits were observed which needs evaluation for selection of superior genotype to use in further crop improvement programme.

Jackfruit trees come up well under humid and warm climate of hill slopes. It is also coming up well in arid and warmer plains of south India. It requires rich deep and well drained alluvial soil or an open textured loamy or lateritic soil supplied with adequate moisture and plant food. Jack is grown upto 1500 MSL but the trees grown above 1200 MSL produce fruits of inferior quality and is very sensitive to frost and drought.

Due to increasing pressure on land and timber value of the jack tree, there has been an unscrupulous felling of jack trees in which many of the outstanding types are lost. Most of the jack types left are also under the threat of felling or becoming extinct. No serious attempts have been made so far on the improvement of the crop. As a preliminary step in crop improvement as well as for commercial exploitation, it is desirable to investigate the nature of variability in terms of vegetative, flowering, fruiting and biochemical characters. Surveys have to be made to identify the elite trees having good agronomic and economic traits with commercial and industrial uses.

A vast area of cashew plantations form in the red lateritic soils of Panruti taluk of Cuddalore district is having jack at intercrop in rows in which a wide range of variability in respect of tree size, bearing habit, fruit size and quality is seen.

Artocarpus Forst. is an unique genus which, on account of its peculiar and edible fruits in certain species, has attracted attention from very remote past. It is a tropical genus with about 50 species (Corner, 1952 and Jarrett, 1959) and native of South East Asia (Barrau, 1976). The species *A. heterophyllus* Lamk. (Syn. *A. integrifolia* L.f. (Syn. *A. intergra* (Thumb.) is native of India and found to be growing wild in Western Ghats (Nairne, 1894 and Gamble, 1972). Jack is basically a diploid 2n =2x=56 (Habib, 1971 and Barrau, 1976).

The Indian species of the genus has got *A. heterophyllus* Lamk, the jack fruit, *A. lakoocha* Roxb, the monkey jack, *A. altilis* (Park) Fosb, the bread fruit, *A. chaplasha* Rowb. and *A. hirsutus* Lamk. (Sharma, 1962). The genus belongs to the tribe Artocarpe, sub family Artocarpoideae which is included in the family Moraceae (Engler and Prantl, 1888; Bailey 1949; Hutchinson, 1959; Rendle, 1971 and Lawrence, 1973). Jarrett (1959) subdivided the genus into *Artocarpus* and *Pseudojaca* based on the position of stipules.

It is widely cultivated in South Asia, the East Indies: Burma, Malaysia, Brazil and other warmer areas of both the hemispheres. Jack trees are distributed in Assam, Bihar, South India and foot hills of Himalayas in North India (Anon, 1948). Jack is a large, evergreen tree of 10-15 m height, indigenous to evergreen forests of the Western Ghats at an altitude of 450 to 1200 m (Anon, 1992).

The tree flowers during December to February. Majority of the female and male inflorescence (catkins) appear on the main trunk and primary branches (Tessy, 1982). It is a highly heterozygous, cross pollinated crop. The fruit is reported to constitute three regions, namely, fruit axis, persistent perianth and true the fruit. The perianth is the most important part and constitutes the major bulk of the fruit. It has basically three regions, namely, lower free and fleshy edible region, middle fused region and the upper free and thorny non-edible region. The fruit consists of four parts, namely (a) flakes (27-40%), derived from fleshy perianth constituting the edible portion, (b) perigones (18-20%), the unfertilized (or) aborted flowers, (c) rind (20-22%) and (d) seeds (20-23%) (Sharma, 1964).

The flakes are used both in unripe and ripe stages. Apart from the table purpose, the ripe flakes can also be used for making canned products, nectar, preserves, jam, jelly, squash, fruit bar and candy. Matured unripe green jack flakes are used for making pickles, chips and papad (Bose and Mitra, 1990).

Variability in Jackfruit

Development of high yielding varieties of crops require information about the nature and magnitude of variability present in the available stock and the selection depends upon a judicious assessment of the available data on phenotypic characters that are connected with yield. Therefore, information on interrelationship between various plant characters with yield and quality among themselves and the extent to which they are influenced by the environment should be known and it has an important bearing for the formulation of an effective breeding plan (Subramanyan and Iyer, 1981).

The aim of exploration or collection is to capture maximum variability of a crop and its wild relatives to meet the needs for conservation. Maintenance of biological diversity helps in sustaining the agricultural productivity (Kamau, 1992)

Jackfruit has innumerable types (or) forms on the basis of the fruit characteristics. The types differ widely among themselves in bearing and density of spines on the rind, size, shape, quality and period of maturity. There is a wide variation in its sweetness, acidity, flavour and taste. These variations among clones offer better scope for its improvement by clonal selection. The types available under various local names have originated in this way (Mitra, 1998).

Jack tree is cultivated in almost all the homesteads of Kerala. Extensive survey has to be undertaken to identify the elite clones which are facing the threat of extinction and steps have to be undertaken to preserve the germplasm (Anon, 1996).

Varieties / types

Cultivated jack fruits are classified into two groups, one with firm flesh and the other with soft flesh. The soft pulp group is locally known as pazham ghula, vela, koozha, ghila or koppa. The mature pericarp of this group is comparatively smaller in size. The juice is either thin or thick, colour varies from pale yellow to dark or golden yellow. Pulp is generally mushy or soft and of varying quality ranging from sweet to insipid. The seeds are comparatively larger. The hard pulp group is locally known as varikka, varcha, kujja, korcha or berka. The pulp is crisp and highly flavoured and therefore relished. The juice is scanty and the seeds are comparatively smaller (Singh *et al,* 1967). Srinivasan (1980) described a variety called, "Muttam varikka" which produced fruits of average weight of 7.0 kg with 46 cm length and 23 cm width.

Bhore *et al.* (1980) identified a highly promising jack type in Ajra village of Kolhapur district of Maharashtra which is high yielding and highly pulpy containing more protein, fat, mineral matter and carbohydrate than local jack fruit. The fruits are small and weighing between 1 and 3 kg and uniform in size and shape.

Jackfruit types like Varikka', TCoozha', 'Navarikka', 'Rudraksha chakka' or Thamara chakka' and other wild forms have been collected from Wynad Plateau in the Western Ghats of Kerala (Anon, 1986). Three types of jackfruits namely 'Rasdar', 'Khajwa' and 'Sugandhi' were identified through survey in the plains of Eastern Uttar Pradesh (Anon, 1988).

Several subgroups are recognized based on taste, shape and size of fruits. One of these is Rudrakshi, a small fruited type, with smooth and less spiny rind and are having a less fleshy perianth than the common jack. There are also certain old trees in Assam which show a tendency to produce fruits during off-season. These are designated as baro-mahia or baro-masha. Ceylon jack and Singapore jack are some introductions in South India. Singapore jack is an early maturing variety and comes to bearing in about 18 months under favourable conditions in low elevation but may take more time at higher elevation. It generally matures in off-season during November to February, in which the fruit is a hard fleshed varikka. In Assam, soft pulp group is divided into six types viz, V1,V2,V3,V4,V5 and V6 and hard pulp group into two varieties, viz, V7 and V8 (Anon, 1992).

In Florida, superior cultivars of jack are Black Gold, Golden Nuggot, J-31, Honey Gold and Lemon Gold (Campbell and McNaughton, 1994).

Burliar-l, Palur-1, and Pechiparai-1 have been released by Tamil Nadu Agricultural University. Through the selection Burliar-1 is a T-Nagar selection developed at Fruit Research Station, Burliar, Palur-1 is a single plant selection isolated in Panikkankuppam village near Panruti and was developed at Vegetable Research Station, Palur. Pechiparai-1 is a selection from Mulagumoodu local developed at Horticultural Research Station, Pechiparai. Hybrid jack was developed at Fruit Research Station, Kallar by crossing Singapore jack and Velippala (Veeraraghavadattam *et al.,* 1996).

Leaf characters

The leaves varied in length from 5.0 to 25.0 cm and in width from 3.5 to 12.0 cm. The shape varied from obviate - elliptic (Anon, 1992).

Maiti *et al,* (2002) reported that there was a significant variation in leaf size and shape. Leaf length varied from 7.7 to 18.7 cm and leaf breadth varied from 6.13 to 11.13 cm.

Fruit characters

The jackfruits from trees of different location / altitude were compared by Hussain and Haque (1977). Average fruit weight ranged from 3.24 to 7.39 kg, the weight of pulp and seed being 0.57 to 0.39 kg, respectively, in the smallest fruit and 2.70 to 1.01 kg, respectively, in the largest fruit. Skin colour ranged on ripening from yellow and pale green to brown. The average fruit weight of jack varied from 3.24 to 17.39 kg (Berry and Kalra, 1988).

A comparative study was carried out on yellow - pulp, light yellow-pulp and orange-pulp types. The light-yellow types had the highest seed weight (7.66 g), seed length (3.23 cm), seed breadth (2.T0 cm), and average total weight of seed per fruit (913.21 g). The yellow types had the more number of seeds per fruit (124.6) and the highest pulp to seed ratio (4.24) (Guruprasad and Thimaraju, 1989). Narasinhan (1990) stated that there were different cultivated types of jack fruit, which varied widely in size, shape, appearance and taste.

From the various genotypes collected in course of surveys of jack fruit growing regions, particularly in Eastern Uttar Pradesh from November 1989 to August 1990, Kumar and Singh (1996) grouped the genotypes into nine categories based on fruit morphology. They observed wide variation in number, shape, size and rind thickness of fruits. Fruit weight ranged from 12.0 to 21.2 kg. All genotypes matured during July-August, except one which matured during March-May. Variability was also noticed in chemical composition of the pulp. The total soluble solids ranged from 14.01 to 25.0 per cent and ascorbic acid content varied from 23.8 to 32.9 mg 100 g^{1}. The best genotype was considered to be AC.7, with moderate yield, large fruits (more than 15 kg) more pulp (about 23 g), small cylinder, high pulp/cylinder ratio and average fruit quality.

A survey was conducted by Sarma *et al.* (1997) during 1995 and 1996 in the lower Brahmaputra valley zone of Assam to study the fruiting behaviour, yield and physico-chemical characters of local jack fruit germplasm. Significant variation was observed in yield and chemical composition of the different genotypes. Fruit yield (60 fruits/tree) and brix values were the highest in KJF 3, while the lowest acid content (6 per cent) was recorded in KFJ 12.

Among 672 trees studied during survey, wide variability has been noticed in yield (14 to 325 fruits/plant), fruit weight (2.1 to 10.2 kg), fruit shape (oblong, roundish and conical), skin colour (greenish, light brown, dark brown and yellowish brown), number of tubercles in skin (5 to 27/cm^2), pulp colour (yellow, whitish, reddish yellow and pinkish yellow), number of segments (34 to380 per fruit), number of seeds (32 to 362 per fruit), texture of pulp (soft, moderately soft and hard), pulp weight (361 to 3648 g), total soluble solids (15.4 to 20.6° brix), total sugars (12.9 to 26.6%) and acidity (0.10 to 0.31%). The growth, bearing (twice a year in some types) and maturity (June to August) of fruits also showed variation among the genotypes (Mitra, 1998).

Singh and Srivastava (2000) reported the existence of similar relationship in jack fruits in different agro-climatic conditions on the basis of physico-chemical characters of fruits. Therefore, these genotypes could be exploited as potential parents for transferring the linear stability for the respective traits to develop high yielding stable genotypes.

Survey, selection, evaluation and conservation of jack fruit genotypes in West Bengal was carried out on the basis of tree characters and physico-chemical characters of fruit under different agro-climatic conditions for the three consecutive years Maiti and Mitra (2002). After studying nearly 2000 trees, 44 genotypes have been identified as superior clones. These clones exhibit significant variations in yield (12 to 1450 fruits), fruit weight (1229 to 17299.75 g), fruit shape, rind colour, flake colour, number of flakes (11 to 425), seed size and shape, texture of flakes, total soluble solids (9.1 to 28.0° brix) total sugars (7.6 to 23.6%), acidity (0.11 to 0.53%) and vitamin C (4.45 to 11.11 mg 100 g^{-1} juice) were observed among the genotypes.

A few jack fruit selections namely NJT- 1, NJT- 2, NJT- 3 and NJT- 4 with large fruits and excellent pulp quality have been identified for table purpose, while types like NJC-1, NJC-2, NJC-3 and NJC-4 were found to be better for culinary purpose (Anon, 2002).

During surveys in the jackfruit growing areas of West Bengal, 16 superior types were identified. A cylindrical shaped type (type -IX) having fruit length of 71 to 80 cm and fruit diameter of 15 to 24 cm weighing 12.5 kg with 263 flakes and 25°brix was identified from Umapur in Mohanpur. Another clone type (type-XV) identified from Barasat has red flakes with TSS of 19.7°brix (Anon, 1992).

Jack fruit *(Artocarpus heterophyllus* Lamk.), a member of the family Moraceae is indigenous to western ghats of India (Santapau, 1966). It can grow under both tropical and subtropical climatic conditions. The fruit of the tree is the largest among the edible fruits and gives more yield than any other fruit tree. It is known as poor man's food in Eastern and Southern parts of India. The fruit is also very popular as vegetable and appears in the market in spring when other vegetables are scarce and expensive. Its fruiting perianths (bulbs) are fine textured having strong sweet aroma and rich appetizing taste at full maturity. Ripe fruits are good source of sugar, minerals and pectin with fair amount of vitamin A. The fruit has several medicinal values. Jacalin, the major protein in jackfruit seeds is a tumeric two chain pectin and suitable for evaluation of the immune status of patients infected with human immunodeficiency virus-l (Sharma *et al.*,2005)

Being cross pollinated and mostly a seed propagated, the jackfruit has enough variability for different morphological and quality characters among the seedling population. The types differ widely among themselves in bearing shape, size, maturity and in quality components such as sugar, minerals, vitamins, acidity, flavour and taste. Such variations offer scope for improvement of Jackfruit through clonal selection.

To initiate any crop improvement programme, studies on genetic divergence is essential for making effective selection of parents for hybridization. Keeping this in view, Rajesh Kannan (2007) studied 30 genotypes to assess the genetic diversity present in the orchards of the Viruthagiri Kuppam and Mudanai Panchayats in Cuddalore District. The salient findings are discussed here under.

Analysis of variance

The analysis of variance revealed significant differences among the 30 genotypes for all the traits of interest. It indicated the existence of sufficient amount of variability among the 30 genotypes of the present study. Therefore, further analysis is appropriate. Similar finding was earlier reported by Maiti *et al.* (2002).

Mean performance

Eleven genotypes exceeded the population mean (+2SE) fruit weight. It was found out that the increase in fruit weight was mainly due to the increase in fruit circumference, fruit rind weight, number of flakes per fruit, weight of flakes with seed, weight of flakes without seed, flake length, flakes width, seed length, seed width and rind flake ratio. The genotype G28 recorded the maximum fruit weight (12.40 kg) and the genotypes Gs registered the lowest mean fruit weight (4.66kg). The genotypes *viz.*, G1, G2, G5, G10, G12, G14, G16, G25, G26, G28 and G30 stand for meritful consideration. These genotypes may be tested for the stability of fruit weight and mass multiplied.

Genetic Diversity

D^2 analysis of the 30 genotypes conformed the presence of high genetic diversity among the types by their resolution into as many as ten clusters. Genotypes of the two reference panchayats were grouped in a single cluster as well as in different clusters (Table 76)

Table 76. Composition of D^2 clusters

Cluster	Numbers of Genotypes	Name of Genotypes.
I	2	MU-7, MU-9
II	5	VR-1, VR-2, VR-3, MU-2, MU-8
III	2	MU-12.MU-14
IV	8	VR-4, VR-5, VR-6, VR-7, VR-8, VR-9, VR-10, MU-3
V	2	VR-12, MU-10
VI	2	VR-13, MU-6
VII	2	MU-l.MU-14
VIII	2	MU-11, MU-13
IX	3	VR-11, VR-14, VR-15
X	2	MU-5, MU-15

Clusters I consisted of two genotypes originating from Mudanai. Cluster II consisted of five genotypes, three hailing from Viruthagiri Kuppam and two originating from Mudanai. Cluster III consisted of two genotypes hailing form Mudanai. Cluster IV consisted eight genotypes, seven originating from Viruthagiri Kuppam and one hailing from Mudanai. Cluster V and VI consisted of two genotypes, one each from Viruthagiri Kuppam and Mudanai. Cluster VII and VIII each consisted two genotypes originating from Mudanai. Cluster IX consisted three genotypes, all hailing from Viruthagiri Kuppam. Clusters X consisted of two genotypes originating from Mudanai.

It is interesting to conceive that the genotypes originating from Mudanai were found grouped along with that of Viruthagiri Kuppam as well as a solo cluster. Similarly, the genotypes hailing from Viruthagiri Kuppam were found grouped along with that of Mudanai as well as solo clusters. This indicated the presence of sufficient amount of genetic diversity among the genotypes of the study. These genotypes could well be exploited in future breeding programme.

Crosses among the divergent presents are likely to yield desirable combinations. Therefore, a crossing programme should be initiated between the genotypes belonging to different clusters based on mean as well as divergent values. In this content, two important points to be considered are i) choice of the particular clusters from which genotypes are to be used as parent in crossing programme and ii) selection of particular genotypes from selected groups.

The greater the distance between two clusters, the wider the genetic diversity among the parents to be included in hybridization programme. Parents combining high yield potential with wider genetic diversity are likely to yield superior plant types (segregants) within short period.

The tendency of genotypes from different panchayats to groups together in one cluster or scattered distribution of genotypes of same origin in different cluster has been observed in the present study. The studies of several workers confirmed, this tendency, that the geographical diversity could not be directly related in any of the tree crops (Nuzhatfaroogui, *et al.,* 1998; Thirugnana Kumar *et al.,* 2006 b). Genetic drift and selection in different environments could cause considerable diversity than geographical distance (Maiti *et al.,* 2002). However, geographical isolation could be important in the alteration of the breeding structure as pointed out by Wallace (1963).

Cluster I recorded minimum intra-cluster value. Cluster I consisted of two genotypes, originating form Mudanai. This may indicate the closeness of the genotypes. Cluster X registered maximum intra-cluster distance. It consisted two genotypes originating from Mudanai. This may indicate that these genotypes were dissimilar, though hailing from same Panchayat (Table 77).

The inter -cluster distance was maximum between the clusters V and VI. Cluster V registered high mean for fruit length, rind weight, number of flakes per fruit and weight of flakes without seed. Cluster VI recorded high mean for flake width, seed length, seed width and hundred seed weight. Cluster V encomposed two genotypes *viz.*, Viruthagiri kuppam-12 and Mudanai-10. Cluster VI composed Viruthagiri Kuppam -13 and Mudanai-6. These genotypes may be crossed among themselves evolve high yielding lines. Rind: Flake ratio followed by fruit weight and hundred seed weight contributed maximum towards genetic diversity (Tables 78 and 79).

As pointed out by Sokal (1965), the choice of characters is important in multivariate analysis and the choice made in the present study appears to be appropriate. Cyclic breeding and crossing may be helpful in bringing new genes into a population of Jack and thus expanding the range of adaption.

It can be concluded that there is no relationship between genetic and geographical diversity in the material under investigation. Inter and intra cluster distance were also not related.

The present investigation was formulated to evaluate the breeding values of Jack genotypes and to identify potential parents for hybridization and selection based on genetic diversity. Thirty genotypes were evaluated for genetic diversity based on 15 fruit characters. The data were subjected to multivariate analysis. The ANOVA indicated significant difference among all the traits of interest. Eleven genotypes *viz*., G1, G2, G5, G10, G12 G14, G16, G25, G26, G28 and G30, registered the highest mean fruit weight. Among these eleven genotypes, G28 stands for merit consideration, as it exhibited the maximum fruit weight. D^2 analysis of 30 genotypes based on 15 fruit characters confirmed the presence of high genetic diversity by their resolution into as many as 10 clusters. The pattern of clustering demonstrated that genetic diversity was not fully related to geographical diversity. Rind: Flake ratio followed by fruit weight and 100 seed weight contributed maximum towards genetic diversity.

Table 77. Average intra and inter cluster D^2 and D values

Cluster	1	II	III	IV	V	VI	VII	VIII	IX	X
1	26.86 (4.88)	279.59 (16.72)	56.97 (7.54)	369.18 (19.21)	568.78 (23.84)	221.02 (14.86)	304.18 (17.44)	497.87 (22.31)	560.66 (23.67)	143.78 (11.99)
II		207.39 (14.40	262.38 (16.19)	250.91 (15.84)	439.42 (20.96)	283.97 (16.85)	198.32 (14.08)	235.55 (15.34)	345.83 (18.59)	250.66 (15.83)
III			54.71 (7.39)	358.36 (18.93)	476.73 (21.83)	309.57 (17.59)	306.52 (17.50)	427.79 (20.68)	448.82 (21.18)	181.29 (13.46)
IV				223.15 (14.93)	460.48 (21.45)	271.63 (16.48)	224.18 (14.97)	274.32 (16.56)	418.52 (20.45)	330.64 (18.18)
V					69.93 (8.36)	805.62 (28.38)	482.52 (21.96)	204.41 (14.29)	336.86 (18.35)	493.77 (22.22)
VI						75.28 (8.67)	225.40 (15.01)	489.52 (22.12)	730.60 (27.03)	209.05 (14.45)
VII							127.72 (11.30)	236.44 (15.37)	385.70 (19.63)	255.38 (15.98)
VIII								146.11 (12.08)	306.26 (17.50)	361.42 (19.01)
IX									248.84 (15.77)	562.03 (23.70)
X										292.32 (17.09)

Table 78. Cluster means for various characters

S. No.	Character	I	II	III	IV	V	VI	VII	VIII	IX	X
1.	Fruit length	37.66	45.93	37.66	45.70	54.66	41.50	45.16	53.66	41.66	43.66
2.	Fruit diameter	28.66	31.93	31.93	30.83	37.66	30.66	33.33	38.83	28.33	33.83
3.	Fruit circumference	87.16	96.86	85.83	97.08	107.50	92.50	88.50	109.50	94.33	96.33
4.	Rind weight	1.96	2.96	1.96	3.20	4.38	2.81	2.73	4.38	2.52	3.25
5.	Number of flakes fruit[1]	82.50	123.20	88.50	120.50	180.00	103.00	129.50	172.33	109.00	119.50
6.	Weight of flakes with seeds	3.25	5.06	3.25	4.69	6.45	4.28	4.85	6.48	4.18	4.58
7.	Weight of flakes without seeds	2.63	4.08	3.25	3.87	5.33	3.26	3.91	5.05	3.63	3.63
8.	Flake length	7.70	7.66	7.43	7.81	7.43	8.28	8.46	7.31	8.06	7.90
9.	Flake width	4.05	4.66	3.51	4.04	3.66	4.31	4.20	3.71	3.91	3.96
10.	Seed length	3.35	3.52	3.20	3.74	3.28	4.00	3.45	3.25	3.26	3.55
11.	Seed width	1.95	2.08	1.95	2.19	1.93	2.33	2.05	1.95	1.93	2.11
12.	Hundred seed weight	804.16	767.66	731.66	791.45	679.16	910.00	813.33	758.33	591.66	836.66
13.	TSS (°Brix)	26.88	20.91	25.68	20.36	25.25	22.01	19.15	19.98	19.97	25.13
14.	Rind: flake ratio	0.60	0.58	0.58	0.72	0.72	0.66	0.60	0.63	0.58	0.60
15.	Fruit weight	5.20	8.04	5.15	7.88	11.05	7.15	7.45	11.13	6.71	8.11

Table 79. Contribution of characters to genetic divergence in jack

S. No.	Characters	Number to first rank	Contribution %
1	Fruit length	1	0.22
2	Fruit diameter	0	0.00
3	Fruit circumstance	0	0.00
4	Rind weight	1	0.22
5	Number of Flakes	10	2.29
6	Weight of Flakes with seeds	0	0.00
7	Weight of Flakes without seeds	0	0.00
8	Flake length	11	2.52
9	Flake width	4	0.91
10	Seed length	16	3.67
11	seed width	2	0.45
12	Hundred seed weight	65	14.94
13	TSS (°Brix)	58	13.33
14	Rind: Flake ratio	138	31.72
15	Fruit weight	129	29.65
Total		435	99.92

11

Genetic Diversity in Gamboge Tree [*Garcinia gummi – Gutta* (L.)]

Thirugnanakumar, S. and K. Inasi

Gamboge tree *Garcinia gummi-gutta* (L.), is a member of the Clusiaceae family. It is a botanical cousin of *Garcinia indica* Choisy. It is a cross pollinated crop. It is called as Kodumpuli in Kerala. It is native of Tropical Asia. It thrives best in the ever green forests of Konkan, stretching from southward to the Kerala coast and Western Ghats upto 1800 m min the Nilgiris (CSIR, 1956). In Kerala, majority of the populations occur naturally in the central districts of Kerala, of different altitudes ranging from sea level to 200 m.

It is one of the potential under-exploited fruit crops. Now, it is gaining momentum for its commercial, industrial and medicinal importance. The dried rind is sold in the market at the rate of Rs. 200 per kg (Muthulakshmi *et al.*, 1999). It is used as a food preservative, spicy source with astringent taste. It is meant for its good flavour, taste and keeping quality. The fruits are too acidic and cannot to eaten raw. The dried rind has unique use in fish curries for imparting its delicate flavour (Sarah, 1998). Dried rind can also be used for polishing gold, silver and coagulating rubber latex. Hydroxy citric acid (HCA) is a very rare natural components being extracted from the rind and is marketed in the name of 'citrin' and 'citrimax'. It has high export value and fetches considerable foreign exchange. In foreign countries, it is used as a herbal medicine against cardiac diseases and for weight management through blocking fat synthesis. The derivatives of acid are potent metabolic regulators of obesity. The unique acid, lowers the blood lipids such as cholesterol and triglycerides by triggering the fatty acid oxidation in the liver through thermogenesis. It mobilizes the body's fat stores and dissolves fat in the liver and also throughout the body paving the way for weight management (Muhammed Majeed, 1994). The 'Gummi-gut' or gamboges is also used as a pigment in miniature paintings, and water colours besides is medicinal use as purgative, hydrazogue and emetic, particularly in dropsies and worm cases. The gum also makes a good varnish. Wood is suitable for match boxes, splints and posts. A knowledge of genetic diversity existing in the population will help to generate a rich base population for initiating proper selection. Measuring genetic diversity in, trees has typically done by using Mahalanobis' D^2 statistic. Genetic diversity plays an important role in tree breeding because hybrids between populations of diverse origin, generally display a greater heterosis than between closely related parents. Studies on genetic diversity are very scarce in Gamboge tree.

The experimental material selected for the present study consisted of 120 one parent families of Gamboge tree *Garcinia gummi-gutta* (L.) which are distributed naturally in the three districts of Kerala *viz.,* Kottayam, Pathanamthitta and Alappuzha. Each tree is considered as a single genotype (Philomina, 2000).

The first phase in tree important program for indigenous species is usually the selection of superior phenotypes from wild populations. Individual tree selection must be based when stands are uneven-aged and when high species diversity makes it impossible to find comparison trees adjacent to the select tree candidate (Beinecke and Lowe, 1969; Ledig, 1974). In the individual tree selection method, the individuals are located and their value for the traits of interest is compared to the average for the altitude, in which the selections are made. The average is a "base line", giving the system its more common name, base line selection. If the candidate exceeds the base line selection by an arbitrary amount, it is selected and incorporated in the breeding population (Cedig, 1974).

One line with the suggestions of Ledig (1974), the breeding populations for the present study were selected and classified as follows.

i. High altitude (Altitude I) - 151 m above MSL
ii. Medium altitude (Altitude II) - 21 m to 150 m above MSL
iii. Low altitude (Altitude III) - Mean sea level to 20 m above MSL

The mean fruit yield of all the trees distributed in a particular altitude was worked out. The individual trees which were above the mean fruit yield +2SE were selected as the reference population for the present study from the initial population of 2000 trees. Accordingly, 40 trees from high altitude, 40 trees from medium altitude and 40 trees low altitude were selected in the first season (June-July 2000). The selected 40 trees from each of the altitudes (totally 120 trees) were studied for their performance in the second season also (June-July 2001).

Thirty fruits from a single tree (as single genotype) were collected and randomly divided into three replications, each consisting of 10 fruits as suggested by Chawhan and Kanwar (2001). Observations were made on X_1) hydroxy citric acid content (%), X_2) length of fruit beak (mm), X_3) weight of the individual fruit (g), X_4) volume of the individual fruit, X_5) density of fruit (g cc^{-1}), X_6) length of fruit (cm), X_8) girth of fruit (cm), X_9) thickness of fruit rind (mm), X_{10}) weight of flesh (g), X_{11}) rind; flesh ratio, X_{12}) weight of dry fruit rind (g), X_{13}) recovery percentage of dry rind (%), X_{14}) number of seeds per fruit and X_{15}) index of seed. The mean values so obtained were subjected to statistical analysis, in a randomized block design, as suggested by Chauhan and Kanwar (2001). All the analysis were done individually for three different altitudes and collectively pooled over different altitudes for two seasons and pooled over seasons.

D^2 analyses of the 40 genotypes in nine analyses and 120 genotypes in three analyses revealed the presence of high genetic diversity among the genotypes by their resolution into seven clusters in Season I Altitude I; six clusters in Season I Altitude II, Season I Altitude III; and Season II Altitude I; five clusters in Season II Altitude II; six clusters II Altitudes III; seven clusters in Seasons pooled Altitudes I; six clusters in Seasons pooled Altitude II; seven clusters in Seasons pooled Altitude III; nine clusters in season I Altitude pooled; eleven clusters in Season II Altitudes pooled and nine clusters in Seasons pooled and Altitude pooled. The genotypes hailing from the same panchayat and altitudes were grouped in into different clusters and the genotypes hailing from different panchayats and altitudes were grouped into the same cluster. Thus, the present study amply indicated the presence of opulence of genetic diversity among the genotypes studied. In fact, the observation of clusters carrying only one genotype, that too from Poonjar (the genotypes *viz.*, G 24 and G 25 of clusters II and III., respectively, in Season I Altitudes I) and Thalavady (G 88 of cluster III, Seasons pooled Altitude III), the confirmed the presence of high diversity among the genotypes studied. These genotypes could well be utilized in heterosis breeding (Tables 80 – 91).

Table 80. Composition of D² clusters and their regional distribution season I and Altitude I

Cluster No.	Total number of genotypes in the cluster	Genotypes coming under each cluster on regional basis
I	11	G1-Koyippuram; G9 and G10-Ranni; G20, G22, G23-Meenachil; G27-Theekkoy; G28; G36, G37 and G38-Melukavu
II	1	G24-Poonjar
III	1	G25-Poonjar
IV	10	G2-Koyippuram; G5-Kozhencherry; G13-Ranni; G17, G19 and G21-Meenachil; G34, G35, G39 and G29-Melukavu
V	5	G15-Meenachil; G32-Melukavu; G26-Theekkoy; G30-Moonnilavu and G6-Kozhencherry
VI	4	G4-Koyippuram; G8, G11 and G12-Ranni
VII	8	G3-Koyppuram; G7-Klozhencherry; G14-Ranni; G16 and G18-Meenachil; G33, G31 and G40-Melakavu

Table 81. Composition of D² clusters and their regional distribution season I and Altitude II

Cluster No.	Total number of genotypes in the cluster	Genotypes coming under each cluster on regional basis
I	11	G41, G44, G46, G52 and G53-Thiruvalla; G58-Peringara; G62 and G64-Mallappally; G69-Poonjar West; G76 and G77-Puliyoor
II	7	G47, G49, G55, G48 and G51-Thiruvall; G57-Peringara; G66-Nattakom East
III	2	G67-Nattakom East; G73-Uzhavoor
IV	13	G42, G45, G43 and G50-Thiruvalla; G60 and G61-Peringara; G65-Mallappally; G 70 and G72-Uzhavoor; G75, G78 and G79-Puliyoor; G80-Chengannoor West
V	5	G54 and G56- Thiruvalla; G63-Mallapapally; G68-Nattakom East and G74-Chengannoor East
VI	2	G59-Peringara; G71-Uzhavoor

Table 82. Composition of D^2 clusters and their regional distribution season I and Altitude III

Cluster No.	Total number of genotypes in the cluster	Genotypes coming under each cluster on regional basis
I	12	G81-Cherthala; G86-Thalavady, G93 and G-94-West Peringara; G116, G114-Aymanam; G119, G120-Udayanapuram; G104-Thiruvarappu; G96, G95-Nedumpuram and G99-Kottayam
II	2	G88-Thalavady and G117-Udayanapuram
III	5	G82 and G84-Cherthala; G103-Thiruvarappu; G91-West Peringara;G109-Aymanam
IV	8	G83-Cherthala; G92-West Peringara; G87-Thalavady; G107-Thiruvarappu; G110, G112 and G115-Aymanam and G98-Kottayam
V	10	G85-Muhamma; G89-Thakazhy; G111, G113 and G108-Aymanam; G100-Kottayam; G102-Chengalam; G105, G106 and G101-Thiruvappu
VI	3	G90-Pallipapadu; G97-Karthikappaly and G118-Udayanapuram

Table 83. Composition of D^2 clusters and their regional distribution season II and Altitude I

Cluster No.	Total number of genotypes in the cluster	Genotypes coming under each cluster on regional basis
I	8	G1, G4-Koyippuram; G9-Ranni; G19, G20-Meenachil; 5-Kozhencherry, G27-Theekkoy and G36-Melukavu
II	5	G11, G12 and G13-Ranni; G26-Theekkoy; G30-Moonnilavu
III	5	G10-Ranni; G22, G23-Meenachil; G24-Poonjar; G28-Melukavu
IV	4	G37, G38-Melukavu; G8-Ranni; G25-Poonjar
V	9	G2, G3-Koyippuram; G32-Melukavu; G6, G7-Kozhencherry; G14-Ranni; G15, G18, G21-Meenachil
VI	9	G16, G17-Meenachil; G29, G39, G31, G33, G40, G34 and G35-Melukavu

Table 84. Composition of D^2 clusters and their regional distribution season II and Altitude II

Cluster No.	Total number of genotypes in the cluster	Genotypes coming under each cluster on regional basis
I	20	G41, G43, G44, G45, G46, G50, G52, G53, G54-Thiruvalla, G62, G63, G64 G72, G76, G77, G78, G79, G75-Puliyoor; G80-Chengannoor; G69-Poonjar West
II	9	G47, G49, G51, G48, G55, G56-Thiruvalla; G57, G58-Peringara; G66-Nattakom
III	2	G67-Nattakom East; G73-Uzhavoor
IV	6	G42-Thiruvalla; G60-Peringara; G65-Mallappally; G68-Nattakom East; G70-Uzhavoor; G74-Chenganoor West
V	3	G59, G61-Peringara; G71-Uzhavoor

Table 85. Composition of D^2 clusters and their regional distribution season II and Altitude III

Cluster No.	Total number of genotypes in the cluster	Genotypes coming under each cluster on regional basis
I	10	G82-Cherthala; G90-Pallippdu; G97-Karthikappally; G98- G100- Kottayam; G101, G106-Thiruvarappu; G113-Aymanam; G118-Udayanapuram; G102-Chengalam
II	6	G85-Muhamma; G89-Thakazhy; G91-West Peringara; G105-Thirucarappu; G108; G111-Aymanam
III	7	G83, G84-Cherthala; G92-West Peringara; G103, G107-Thiruvarappu;G110, G112-Aymanam
IV	4	G93-West Peringara; G109, G114-Aymanam; G119-Udayanapuram
V	9	G86-G87-Thalavady; G94-West Peringara; G97; G96-Nedumpuram; G99-Kottayam; G115 G116-Aymanam; G120-Udayanapuram
VI	4	G81-Cherthala; G88-Thalavady; G104-Thiruvarappu G117-Udayanapuram

Table 86. Composition of D^2 clusters and their regional distribution season pooled Altitude I

Cluster No.	Total number of genotypes in the cluster	Genotypes coming under each cluster on regional basis
I	10	G1-Koyippuram; G10-Ranni; G20, G22, G23-Meenachil G27-Theekkoy, G28, G36 G37, G38-Melukavu
II	1	G24-Poonjar
III	1	G25-Poonjar
IV	8	G4-Koyippuram; G8, G13, G9, G11, G12-Ranni; G26-Theekkoy; G30-Moonnilavu
V	12	G2-Koyippuram; G5-Kozhencherry; G15, G17 G19-Meenachi; G29,G31, G32, G33, G34, G35 G39-Melukavu
VI	8	G3-Koyippuram; G7, G6-Kozhencherry; G14-Ranni; G18 G21-Meenachil
VII	2	G16-Meenachil; G40-Melukavu

Table 87. Composition of D^2 clusters and their regional distribution season pooled Altitude II

Cluster No.	Total number of genotypes in the cluster	Genotypes coming under each cluster on regional basis
I	15	G41, G44, G46, G48, G47, G49, G51, G52-Thiruvalla G57 G58-Peringara; G64-Mallappally;G66-Nattakom East G77G79-Puliyoor; G80-Chengannoor
II	3	G55, G56-Thiruvalla; G62-Mallappally
III	9	G43, G50, G53, G54-Thiruvalla; G57, G76, G78- Puliyoor G72-Uzhavoor; G69-Poonjar West
IV	2	G67-Nattakom East; G73-Uzhavoor
V	9	G42, G45-Thiruvalla; G74-Chenganoor West; G59, G60-G61-Peringara; G63-Mallappally; G68-Nattakom East G70- Uzhavoor
VI	2	G59-Peringara; G71-Uzhavoor

Table 88. Composition of D^2 clusters and their regional distribution season pooled Altitude III

Cluster No.	Total number of genotypes in the cluster	Genotypes coming under each cluster on regional basis
I	11	G81-Cherthala; G86-Thalavady; G94-West Peringara; G95-Nedumpuram; G99-Kottayam; G104- Thiruvarappu; G114 G115-Aymanam; G117, G119 and G120- Udayanapuram
II	3	G96-Nedumpuram; G87-Thalavady and G116-Aymanam
III	1	G88-Thalavady
IV	6	G82-Cherthala; G90-Pallippadu; G98-Kottayam; G101- G103-Thiruvaprappu; G102-Chengalam
V	8	G83, G84-Cherthala; G92, G93-West Peringara; G107-Thiruvarappu; G109, G110, G112-Aymanam
VI	4	G85-Muhamma; G91-West Peringara; G100-Kottayam; G111-Aymanam
VII	7	G89-Thakazhy; G97-Karthikappally; G105;G106 Thiruvarappu G108, G113-Aymanam; G118-Udayanapuram

Table 89. Composition of D^2 clusters and their regional distribution season I Altitudes pooled

Cluster No.	Total number of genotypes in the cluster	Genotypes coming under each cluster on regional basis
I	32	G1, G4-Koyippuram; G36-G28-Melukavu; G20, G22-Meenachil; G94-West Perngara; G99-Kottayam; G27-Theekkoy G56, G42, G45-Thiruvalla; G119, G120-Udayanapuram; G86-Thalavady; G5-Kozhencherry; G60, G61-peringara; G104-Thiruvarappu G70-Uzhavoor; G63-Mallappally; G81-Cherthala; G74-Chengannoor West; G95, G96-Nedumpuram; G114, G116-Aymanam; G68-Nattakom East; G8, G9, G10, G11-Ranni
II	10	G37, G38-Melukavu, G23-Meenachil; G88-Thalavadi; G59-Peringara; G65-Mallappally; G117-Udayanapuram; G77-Uzhavoor; G25, G24-Poonjar

III	20	G2-Koyippuram; G29, G32-Mukukavu, G15, G17-Meenachil; G26-Theekkoy; G43, G50-Thiruvalla; G72-zhavoor; G82, G84-Cherthala G30-Moonilavu; G109, G10, G112-Aymanam; G12, G13-Ranni; G75-Puliyoor; G93-West Peringara; G103-Thiruvarappu
IV	3	G55-Thiruvalla; G87-Thalavady; G107-Thiruvarappu
V	6	G3-Koyippuram; G18-Meenachil; G6-Kozhencherry; G73-Uzhavoor; G64-Mallappally; G67-Nattakom East
VI	10	G33-Melukavu; G7-Kozhencherry; G111-Aymanam; G44, G53, G54-Thiruvalla; G14-Ranni; G91-Peringara; G76-Puliyoor; G69-Poonjar West
VII	18	G34, G35, G, 39, G31-Melukavu, G77, G78-Puliyoor; G19, G21-Meenachil; G98-Kottayam; G41, G46, G52-Thiruvall; G115-Aymanam; G106, G101, G105-Thiruvarappu; G102-Chengalam; G85-Muhamma
VIII	6	G83-Cherthala; G92-West Peringara; G97-Karthikappally; G80-Chengannoor West; G79-Puliyoor; G90-Pallippadu
IX	15	G40-Melukavu, G16-Meenachil; G89-Thakazhy; G47, G48, G49, G51-Thiruvalla; G57, G58-Peringara; G62-Mallappally; G108, G113-Aymanam; G118-Udayanapuram; G66-Nattakom East; G100-Kottayam

Table 90. Composition of D[2] clusters and their regional distribution season II Altitudes pooled

Cluster No.	Total number of genotypes in the cluster	Genotypes coming under each cluster on regional basis
I	18	G1-Koyippuram; G26-Theekkoy; G63-Mallappally; G50, G43, G44-Thiruvalla; G83 G84-Cherthala; G92-West Peringara; G110, G112, G109-Aymanam; G107-Thiruvarappu; G30-Moonnilavu
II	24	G2, G3-Koyippuram; G6, G5-Kozhencherry; G32 G35, G33, G31, G34, G40-Melukavu; G58-Peringara; G15, G16, G21, G19-Meenachil; G45 G46, G52-Thiruvalla; G14-Ranni; G106-Thiruvarappu G77-Puliyoor; G64, G62-Mallappally; G113-Aymanam
III	7	G18-Meenachil; G54-Thiruvalla; G7-Kozhencherry; G73, G72-Uzhavoor; G91-West Peringara; G67-Nattakom East
IV	8	G48, G55, G56-Thiruvalla; G87-Thalavady G108-Aymanam; G93-West Peringara G105-Thiruvarappu; G89-Thakazhy
V	19	G29,G39-Melukavu, G80-Chengannoor; G17-Meenachil; G103, G101-Thiruvarappu; G41, G49 G53, G51, G47-Thiruvalla G102-Chengalam; G82-Cherthala; G98-Kottayam G69-Poonjar West; G75, G76, G78-Puliyoor G90-Pallippadu
VI	6	G57-Peringara; G111-Aymanam; G66-Nattakom East; G100-Kottayam; G79-Puliyoor; G85-Muhamma
VII	2	G97-Karthikappally; G118-Udayanapuram
VIII	4	G4-Koyippuram; G114-Aymanam; G9-Ranni G68-Nattakom East
IX	17	G36, G28-Melukavu; G27-Theekkoy; G20, G22-Meenachil; G86-Thalavadi; G60-Peringara; G70-Uzhavoor G10-Ranni; G95, G96-Nedumpuram; G94-West Peringara; G115, G116-Aymanam; G99-Kottayam G120-Udayanapuram; G74-Chengannoor West

X	12	G37, G38-Melukavu, G23-Meenachil; G42-Thiruvalla; G88-Thalavadi; G61-Peringara; G65-Mallappally; G81-Cherthala; G8-Ranni; G117-Udayanapuram; G104-Thiruvarappu; G24-Poonjar
XI	3	G59-Perngara; G71-Uzhavoor; G25-Poonjar

Table 91. Composition of D² clusters and their regional distribution season and Altitudes pooled.

Cluster No.	Total number of genotypes in the cluster	Genotypes coming under each cluster on regional basis
I	27	G1-Koyippuram; G20, G22-Meenachil; G42-Thiruvalla; G99-Kottayam; G36, G37, G38, G28-Melukavu; G27-Theekkoy; G86; G88-Thalavady; G24-Poonjar; G61-Peringara; G65-Mallappally; G81-Cherthala; G8, G10-Ranni; G95, G96-Nedumpuram; G115, G116-Aymanam; G94-West Peringaram; G117, G120-Udayanapuram; G104-Thiruvarappu; G74-Chengannoor West
II	4	G23-Meenachil; G59-Peringara; G71-Uzhavoor; G25-Poonjar
III	18	G2-Koyippuram; G29-Melukavu; G17-Meenachil; G43, G55, G50, G47-Thiruvalla; G83-Cherthala; G110, G112-Aymanam; G13-Ranni; G92-West Peringara; G101, G103, G107-Thiruvarappu; G98-Poonjar; G75-Puliyoor
IV	11	G32, G33, G34, G35-Melukavu, G15, G19-Meenachil; G45, G56-Thiruvalla; G62-Mallappaly; G93-West Peringara; G106-Thiruvarappu
V	15	G4-Koyippuram;G26-Theekkoy; G5-Kozhencherry; G60-Peringara; G70-Uzhavoor; G63-Mallappally; G84-Cherthala; G119-Udayanapuram; G109, G114-Aymanam; G68-Nattakom East; G9, G11, G12-Ranni; G30-Moonilavu
VI	8	G3-Theekkoy, G18, G21-Meenachil; G6-Kozhencherry; G73-Uzhavoor; G64-Mallappally; G67-Nattakom East; G77-Puliyoor

VII	19	G39-Melukavu, G9-Kozhencherry; G52, G41, G44, G53, G46, G54-Thiruvalla, G90-Pallippadu; G72-Uzhavoor; G82-Cherthala; G69-Poonjar; G14-Ranni; G91-West Peringara; G102-Chengalam; G80-Chengannoor West; G76, G78, G79-Puliyoor
VIII	7	G40, G31-Melukavu, G16-Meenachil; G58-Peringara; G113-Aymanam; G118-Udayanapuram; G97-Karthikappally
IX	11	G48, G49, G51-Thiruvalla; G57-Peringara; G85-Muhamma; G108, G111-Aymanam; G89-Thakazhy; G105-Thiruvarappu; G66-Nakkatom East; G100-Kottayam

The tendency of genotypes from diverse geographic regions to group together in one and the same cluster or scattered distribution of genotypes of same geographic origin in different clusters has been observed in the present study. The studies of several workers have confirmed this tendency, that the geographical distribution and genetic diversity could not be directly related in any of the crop, tobacco (Murthy, 1965), linseed, wheat and sorghum (Murthy and Arunachalam, 1966), cotton (Singh and Gupta, 1968); safflower (Ranga Rao *et al.,* 1980) and sesame (Thirugnana Kumar, 1991). This is not surprising since *Garcinia gummi-gutta* (L.) is a cross pollinating tree crop and hence free exchange of genetic material might have occurred and occurring in these reference populations. In fact, the geographical location, the direction of wind and human civilization supports the above fact.

The present study have shown that factors other than geographic divergence might have been responsible for the differential grouping of genotypes. This might have been due to the reasons that biotypes originating in a particular habitat have utility only for certain characters for which selection has been practiced. Therefore, varieties originating at the same panchayat or altitude might have different architecture. For example, the genotypes originating from Meenachil were grouped in Cluster I and II in seasons pooled and altitude pooled analysis. Likewise certain genotypes might possess similar characteristics even though their origin were from different panchayat or altitude. For example, the Cluster I of seasons pooled and altitudes pooled analysis composed of genotypes originating from Meenachil, Thalavady, Aymanam, Ranni and Udhayanapuram. Therefore, genotypes with same geographic origin could have undergone changes in the characteristics under selection for different characters. Thus, the clustering pattern failed to indicate any relationship between genetic divergence and geographical distribution.

On the other hand, genetic drift, and selection in different altitudes could have caused greater genetic diversity in the reference population than the geographical distance. Murthy *et al.* (1962) suggested genetic drift and local selection as causes of sub-specific different in *Nicotiana rustica.*

In the certain cases, the effect of geographic origin appeared to have influenced the clustering pattern. For example, the genotypes *viz.,* G24 and G25, originating from Poonjar panchayat were grouped separately in Season I Altitude I and Seasons pooled Altitude I analyses. Similarly, the genotype *viz.,* G88, originating from Thalavady, was grouped separately in Seasons pooled and Altitude III analysis. Such an effect of geographic origin in clustering pattern has been recorded by Jeswani *et al.* (1970) in linseed; Ziauddin Ahmad *et al.* (1980) in *Triticale* and Thirugnanakumar (1991) in sesame. It is suggested that, although geographical distribution could not be taken as a sole criterion of genetic diversity in *Garcinia gummi-gutta* (L.) the influence of the former could still be traced.

12

Molecular Genetic Diversity in Rice [*Oryza sativa* (L.)]

Anandan, A. and S. Thirugnanakumar

Rice (*Oryza sativa* L.) is the most popular crop and source of food for more than one-third of the world's population. About 90 % of the world's rice is grown in Asia and more than 3 billion Asians rely on rice for 30–75 % of their total calories (Braun and Bos 2004). It has been cultivated in four different ecosystems viz., irrigated, rainfed low land, upland and flood-prone (Halwart and Gupta, 2004). Since rice planting has been started around 8,000 to 9,000 years ago, the level of diversity would be more than it expected. The genetic diversity plays a major role in survival and adaptability of a species, as changes happen in environment. Organism needs to make changes in the phenotype that enables it to adapt and survive in unfavorable conditions. The inadequate knowledge on diversity of a species reduces the chance of widening the genetic base of that species. The venture of green revolution in Indian agriculture has narrowed down the genetic base of any popular crops. Therefore, collection, conservation and utilization of germplasm are very much needed in the present scenario for improving the varieties against biotic and abiotic stress.

Morphological and biochemical trait based diversity studies being largely influenced by environment as phenotype and/or enzymes are altered by the environmental and developmental changes. Therefore, adapting DNA based molecular markers for genetic diversity studies would be more reliable to overcome the environmental influence. Being molecular markers as a versatile tool, it establishes its strong holds in several fields of life science. Molecular marker based technique does not require DNA sequencing but it is generally based on nucleic acid hybridization or polymerase chain reaction (PCR). The PCR based tests has several advantages over hybridization technique as it is cheaper, speedy and sensitivity. Therefore, PCR based markers have been used extensively in genetic diversity study for assessing genetic variation within the species. The recent advent of molecular and computational tools now enables the estimation of genetic diversity and allele frequency between the genotypes easy. In this chapter 91 rice accessions were studied for their genetic diversity using 35 polymorphic SSR markers. The details of molecular markers information were collected in public domain database www.gramene.org.

Molecular data analysis

The molecular data from 35 polymorphic SSR markers were analyzed using specific statistical software to determine relationship between the diverse rice accessions. The genetic diversity parameters such as number of alleles per locus, allele frequency, heterozygosity and polymorphic information content (PIC) were estimated using the program POWERMARKER Ver3.25 (Liu and Muse 2005). Allele frequency represents the frequency of particular allele for each marker. Heterozygosity is the proportion of heterozygous individuals in the population. Polymorphic information content that represent the amount of polymorphism within a population was estimated based on Botstein et al. (1980). The PIC values for each SSR were estimated by determining the frequency of alleles per locus using the following formula:

$$PIC = 1-\sum x_i^2$$

Where, x_i is the relative frequency of the ith allele of the SSR loci.

Further, the allelic data were subjected to estimation of genetic distances among genotypes using simple matching coefficients by bootstrapping 10,000 times and they were clustered using neighbor joining method (Felsenstien 1985). The gene diversity analysis provides an unbiased estimation of genetic variation at any given locus. The clustering analysis was done using DARwin software ver 5.0 (Perrier *et al.* 2003; Perrier and Jacquemoud-Collet 2006).

Interpretation

Genetic variability of 91 accessions of rice was evaluated for the estimation of genetic diversity parameters. Of the total of 49 SSR markers used for genotyping of rice accessions, 35 markers showed polymorphism and were used in further analyses. The remaining 14 markers were not polymorphic and hence were excluded. The 35 markers generated a total of 82 distinct alleles across 91 rice genotypes with an average of 6 alleles per locus.

The number of alleles varied from 1- 4 with an average number of alleles amplified per marker was 2.34. The maximum numbers of four alleles were amplified with RM336 and three alleles each were amplified with ten markers (RM3839, RM161, RM9, RM148, RM340, RM16, RM252, RM8085, RM106 and RM341). The variability in the number of alleles identified per locus might be due to utilization of diverse rice genotypes. The size of the amplified fragments ranged from 70bp to 750bp. The PIC value of the polymorphic primers RM161, RM7075, RM348 and RM106 showed more than 0.374 PIC value, while in RM223 exhibited the lowest PIC value of 0.075 with an average PIC value of 0.32. Details of the polymorphic SSR markers are given in Table 92. Out of 35 SSRs, RM230 didn't show any heterozygosity and in the remaining of SSR markers ranged from 0.0132 (RM263) to 0.8235 (RM161) with an average of 0.271 and gene diversity ranged from 0.0781 (RM223) to 0.4994 (RM161) with an average of 0.425. The gel images of amplified fragment with polymorphic markers are represented in Fig.3. The major allele frequency (MAF) at the SSR loci varied between 0.517 (RM161) and 0.959 (RM223), with an average value of 0.649 (Table 92) of the 91 rice genotypes shared common allele per locus.

Table 92. Details of SSR molecular marker loci used for genotyping of 91 rice accessions and their genetic diversity parameters as major allele frequency (MAF), total number of alleles per locus, gene diversity, heterozygosity and polymorphism information content (PIC) .

S.No	Markers	Major Allele Frequency	Total No. of Allele	Gene Diversity	Heterozygosity	Polymorphic Information Content (Pic)
1	RM13	0.5846	2	0.4857	0.0308	0.3677
2	RM125	0.9231	2	0.1420	0.1538	0.1319
3	RM26	0.6689	2	0.4429	0.4730	0.3448
4	RM87	0.8103	2	0.3074	0.3333	0.2601
5	RM3839	0.9397	3	0.1134	0.0172	0.1070
6	RM161	0.5176	3	0.4994	0.8235	0.3747
7	RM218	0.7500	2	0.3750	0.2778	0.3047
8	RM9	0.6111	3	0.4753	0.5833	0.3623
9	RM7075	0.5309	2	0.4981	0.2716	0.3740
10	RM258	0.6220	2	0.4703	0.0488	0.3597
11	RM3428	0.5298	2	0.4982	0.0595	0.3741
12	RM334	0.5984	2	0.4807	0.0164	0.3651
13	RM148	0.7013	3	0.4190	0.5714	0.3312
14	RM221	0.9143	2	0.1567	0.1429	0.1445
15	RM264	0.5542	2	0.4941	0.0482	0.3720
16	RM230	0.6447	2	0.4581	0.0000	0.3532
17	RM85	0.6575	2	0.4504	0.0548	0.3490
18	RM340	0.5357	3	0.4974	0.2857	0.3737

19	RM249	0.6023	2	0.4791	0.5682	0.3643
20	RM225	0.7188	2	0.4043	0.0875	0.3226
21	RM263	0.5329	2	0.4978	0.0132	0.3739
22	RM223	0.9593	2	0.0781	0.0814	0.0750
23	RM1339	0.6964	2	0.4228	0.0357	0.3334
24	RM16	0.6076	3	0.4768	0.5823	0.3632
25	RM252	0.5949	3	0.4820	0.7595	0.3658
26	RM336	0.6000	4	0.4800	0.7333	0.3648
27	RM8085	0.5706	3	0.4900	0.1294	0.3700
28	RM19	0.7143	2	0.4082	0.0952	0.3249
29	RM106	0.5230	3	0.4989	0.4943	0.3745
30	RM168	0.5432	2	0.4963	0.4444	0.3731
31	RM224	0.5345	2	0.4976	0.2759	0.3738
32	RM7389	0.6528	2	0.4533	0.3611	0.3506
33	RM21	0.5833	2	0.4861	0.2667	0.3680
34	RM341	0.5857	3	0.4853	0.3429	0.3675
35	RM253	0.6027	2	0.4789	0.0274	0.3642
	MEAN	0.6490	2.34	0.4251	0.2712	0.3280

Cluster based analysis

The Cluster analysis was carried out with 35 SSR markers to assess genetic distance of 91 rice genotypes and the dissimilarity matrix-using neighbor joining method. In the Unrooted tree, genotypes were grouped into 11 distinct clusters (Fig. 4). The identified clusters were again sub-clustered into 21 based on the marker data analysis through POWERMARKER software. The 91 genotypes showed wider genetic variation between genotypes. Among the total 11 major clusters, higher rice accessions contained in Cluster II with 25 rice genotypes (Abhisek, B15, B6, CR Dhan 40, B12, Neela, B17, Sekri, Heera, B23, B21, B13, B22, B20, B18, B19, AC 100296, AC 100010, Boff16, AC 100193, Kabuk Phou(A), CR Dhan 103, Vandana). The lowest number of rice accessions were observed in Cluster VI, consists of two rice varieties (Brown gora and Harishankar(B)) (Table 93). However, Cluster VII to XI contained with one accession each (AC 100107, Sada Bahar, Gini, AC 100209(B), AC 100326) were showed mono-genotypic clustering pattern (Fig.4).

The genotypes from different clusters may be chosen for hybridization. The genotypes from extreme to moderate distantly placed genotypes could be hybridized to exploit more heterosis and superior segregants.

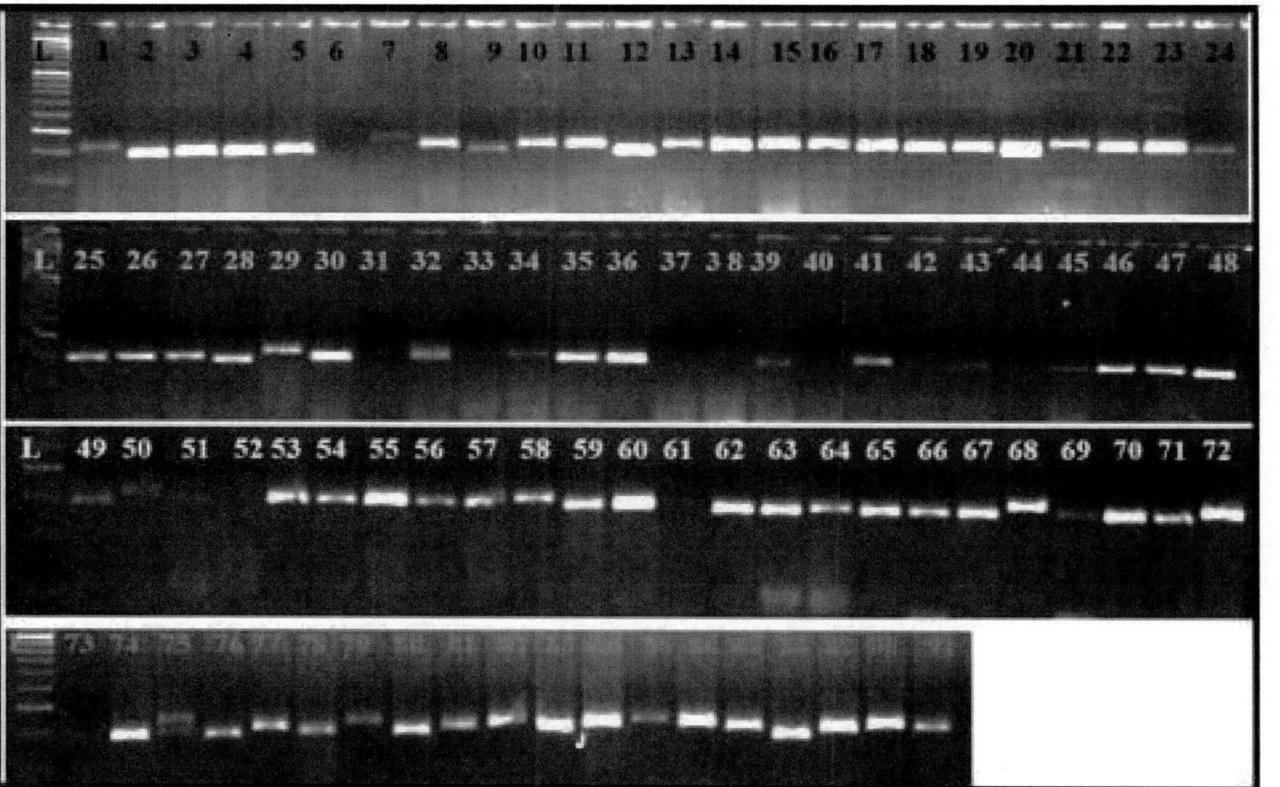

L-50bp ladder; 1-91 rice genotypes; RM-Rice Microsatellite

Fig. 3: A representative gel picture of RM3428 marker with 91 rice accessions

Table 93: Distribution of rice genotypes into different clusters

S.NO	CLUSTERS	NO. OF GENOTYPES/ CLUSTER	SUB-CLUSTERS	NO. OF GENOTYPES/ SUB-CLUSTER	NAME OF THE VARIETIES/ACCESSIONS
1	I	15	2	8	SUKHA PANKI, DULAR, CHAKHAO AUBI, B8, B10, AC100142 SARATHI, B11
				7	BULU HARANA, LONG MANABI (B), AKHIYA TURFA, ARUPATHAM KURUVAI(B), LONG MANABI (A), KUMBHI PHOU, ARUPATHAM KURUVAI (A)
2	II	25	4	11	B16, ABHISEK, B15, CR DHAN 40, KASALATH, B12, NEELA B17, SEKRI, AC00296, HEERA
				9	B23, B21, B13, AC100006, B22, BOFF16, B20, B18, B19
				3	AC100193, KABUK PHOU(A), B1
				2	CR DHAN 103, VANDANA
3	III	12	2	6	AC100062(A), AC100062(B), AC100062 (C), KALINGA – 3, KABUK PHOU(B), AC100124
				6	AC100175, B6, AC100133, AC100026, AC100121, AC100123
4	IV	11	2	7	AC100219(B), AC100281, PHALGUNI(A), LEIMA PHOU, AC100169, IR36, KAMESH
				4	AC100142, AC100219(A), KHITISH, PHALGUNI (B)

5	V	21	5	7	AC100329, AC100337(B), AC100203, AC100032(A) AC100284, AC100281, AC100295
				6	AC100015, AC100282(A), AC100301, AC100309, AC100135, AC100328
				3	AC 100087, AC 100032(A), AC 100035
				3	AC 100285(B), HARISHANKAR(A), AC 100283
				2	BAMAN PHOU, AC 100010
6	VI	2	1	2	BROWN GORA, HARISHANKAR(B)
7	VII (MONOGENOTYPIC CLUSTER)				GINI
8	VIII (MONOGENOTYPIC CLUSTER)				AC 100209(B)
9	IX (MONOGENOTYPIC CLUSTER)				SADA BAHAR
10	X (MONOGENOTYPIC CLUSTER)				AC 100326
11	XI (MONOGENOTYPIC CLUSTER)				AC 100107

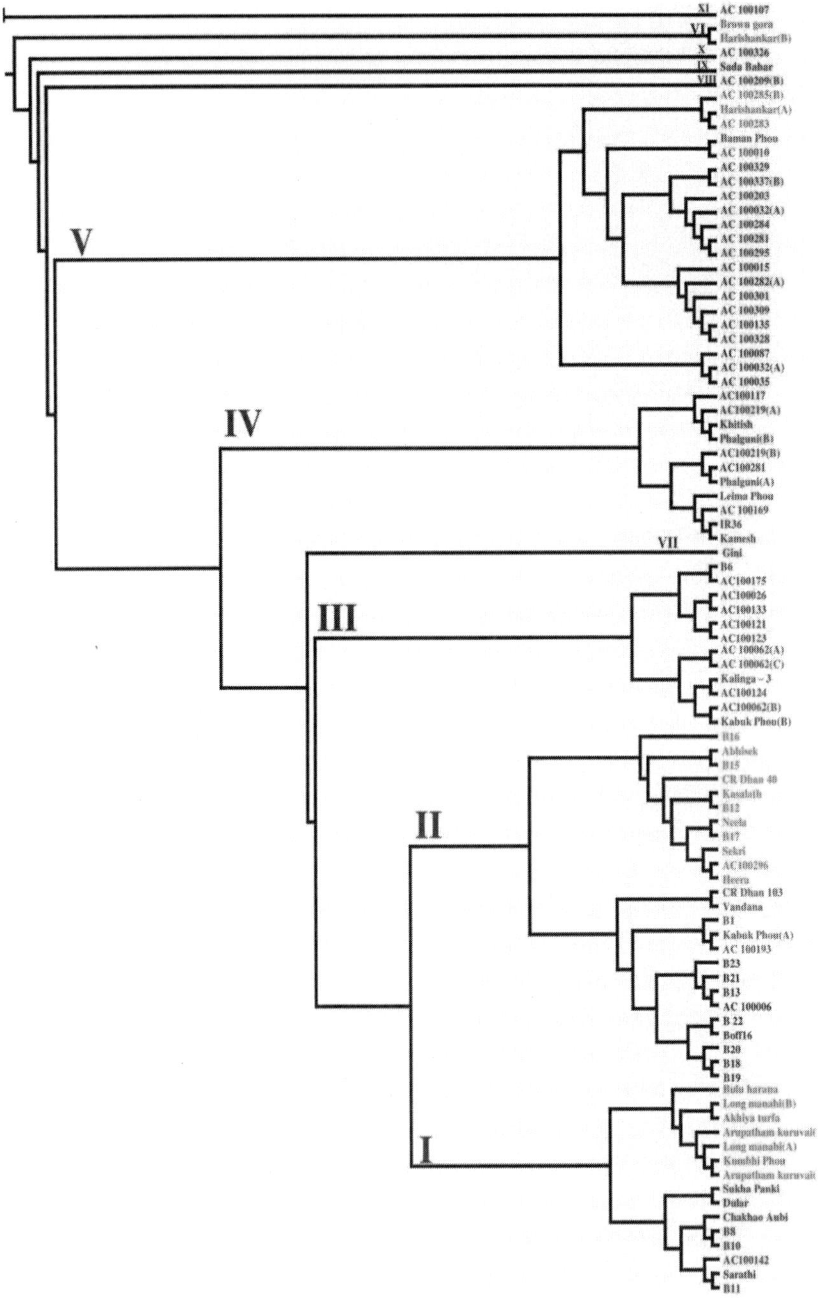

Fig. 4: Neighbour-joining tree showing the genetic relationships among the 91 genotypes rice of accessions based on SSR polymorphic marker data

13

Stability Analysis in Crop Plants

A successfully developed new cultivar should have stable performance and broad adaptation over a range of environments, seasons and locations, in addition to high yield potential. Evaluating performance stability and range of adaptations is becoming increasingly important in breeding programs. Since stability and adaptability are important selection criteria in breeding programs, in depth research on stability is needed for a better estimate of crop performance and adaptability (Lin and Binns, 1988).

Having realized the importance of genotype × environment, several biometrical models have been developed to analyse the stability of genotypes. Such models have been discussed, critically reviewed comprehensively by Knight (1970), Freeman (1973), Hill (1975), Brain (1986) and Gautam *et al.* (1986).

However, the various methods proposed for the statistical analysis of genotype × environment are based on one of following approaches (Jain, 1982). In the first the variance interactions are estimated by equating the observed mean squares in the analysis of variance to their expectations on the random model. The second approach called phenotypic approach by Hill (1975) used the technique of partitioning the genotype × environment component of variability into its linear and non-linear portions of assessing the stability of genotypes over a range of environments. This approach which is very widely adopted in plant breeding, came to be known as "Joint regression analysis".

Many workers contributed for the development of this approach, notably Finlay and Wilkinson (1963), Eberhart and Russell (1966), Freeman and Perkins (1971) and Shukla (1972). The third approach called genotypic approach which also suggested by Hill (1975) is based on fitting of general models in which it specify, the contribution of genetic, environmental and genotype × environment to the generation means and variance. The contributions for he development of this approach were from Bucio (1966), Bucio and Hill (1966),

Bucio *et al.* (1969), Perkins and Jinks (1968), Breese (1969), Hill and Perkins (1969) and Perkins (1970). Linear regression and joint regression models were more commonly used due to their simplicity and reliability in biological experiments. In general, these approaches fall conveniently into two parts.

i. Conventional analysis of variance to know the existence of significant genotype × environment and then proceed to
ii. Joint regression analysis to determine, whether or not genotype × environment interactions are a linear function of the additive environmental component (Hill, 1975).

According to Verma and Virk (1978) non-independence of environmental value of phenotypic variables and non-linearity of response of genotype to environmental value, are the drawbacks of the regression models. Independent environmental values has been suggested to be obtained from

i. Replicating same material into two groups, one for measuring genotype ′ environment and the other to assess environmental index.
ii. Inclusion of limited number of assessment genotype with test genotypes.
iii. Physical measures
iv. Using a set of genotypes closely related to the trial genotypes

In addition to this important criticism, there are a few more criticisms expressed by other authors such as Baker (1969) and Byth *et al.* (1986). They criticized the linear regression approach on the grounds that in an ANOVA the proportion of genotype × environment contributable to linear regression on the environmental indices may be very small. According to Verma and Chahal (1978), conventional regression approach cannot detect the theoretically ideal genotype, which has been defined as the one with relatively low sensitivity in the poor environment and high sensitivity in the favourable environment. Brennan and Byth (1979) stated that linear regression analysis is valid only when linearity is preponderant. Brian (1986) opined that in the linear regression model, stability statistic of a variety may be unduly influenced by its performance in only one or two environments and to that extent may be seriously misleading.

Principle component analysis (Mahajan and Prasad, 1986) helps to provide a two dimensional representation of genotype × environment pattern which allows the response of each variety in each environment to be directly identified.

Apart from linear and joint regression models, there are several models and approaches proposed by several workers, which have been reviewed by Brian (1986). However, these models have not been used extensively in biological experiments.

Different workers based on stability parameters considered and defined stable genotypes in different ways. Lewis (1954) defined stability factor (S.F.) for the ith genotype as S.F. = XHE/XLE, where X is mean, HE and LE are high and low yielding environments respectively. A unit value of S.F. indicates maximum phenotypic stability in this computation. Wricke (1962) considered 'ecovalence' which is the contribution of a genotype to the genotype × environment, sum of squares as a stability parameter. Plaisted and Peterson (1959) proposed similar parameter.

According to Finlay and Wilkinson (1963), the stability parameter of a genotype is its phenotypic regression co-efficient (b_i), a measure of the response of the ith variety to changing environments. The genotype with a unit 'b_i' value and higher mean yield (X_i) is said to be stable variety for a range of environments. As the mean yield decreases, genotypes with high or low slopes are regarded as being specifically adopted to favourable and unfavourable environments respectively.

Eberhart and Russell (1966) who extend the model of Finlay and Wilkinson (1963), regarded deviation from regression of ith variety at j^{th} environment (d_{ij}) as another important component of varietal stability, a stable variety being one with a regression slope near to zero deviation from regression line and higher mean yield.

Ram *et al.* (1970) partitioned phenotypic index (P_i) as an easier method of estimating the phenotypic stability. Phenotypic index of a genotype is the average stability sum of deviations of the genetic mean performance and the grand mean.

Tai (1971) and Shukla (1972) partitioned the genotype × environment sums of squares into components for each genotype separately by considering stability variance over environments of (g_{ij}–e_{ij}) where g_{ij} is interaction between i^{th} genotype at j^{th} environment and e_{ij} is the mean of error component over replications.

Choudhary *et al.* (1972) suggested a method to work out an adaptability index (A_i) as a measure of stability. The magnitude of 'A_i' indicated degree of adaptability and its sign determined the nature of performance. Bilbro and Ray (1976) used the regression co-efficients (b_i) of yields on environment indices as a measure of adaptation and the coefficient of determination (r^2) as a stability measure.

Luthra and Singh (1974) and Verma and Virk (1978) compared some stability models and parameters and concluded that the relative rankings of the genotypes in Eberhart and Russell's and Perkin and Jink's model would be same.

Stability parameters

Pooled analysis of variance over seasons

The ANOVA is constructed as indicated herein

Source	D.F.	M.S.	Expectations of mean squares
Block within season	1 (b-1)	-	-
Season (1)	(1-1)	-	-
Genotype (g)	(g-1)	M_g	$\sigma_e^2 + b\sigma_{gl}^2 + bl\,\sigma_g^2$
Genotype × Season	(g-1) (1-1)	Mg_l	$\sigma_e^2 + b\sigma_{gl}^2 +$
Pooled error	n(b-1) (g-1)	M_e	σ_e^2

The $\square_e^2$ (genotypic component) and $\square_{gl}^2$ (genotype × season interaction) estimates are obtained as follow:

$$\sigma_e^2 = \frac{M_g - M_{gl}}{bl}$$

$$\sigma_{gl}^2 = \frac{M_{gl} - M_e}{b}$$

Where b = Number of blocks

1 = Number of sesame

Stability parameters

Following the methodology of Eberhart and Russell (1966) three parameters of stability namely mean, regression coefficient 'b' and mean squared deviation (s_d^2) are estimated for each genotype.

The linear model proposed by Eberhart and Russell (1966) is

$$Y_{ij} = \square_i + b_i I_j + \square_{ij}$$

Where,

Y_{ij} = Mean performance of ith genotype in j^{th} season

$\square_i$ = Average performance of ith genotype over all seasons

b_i = Regression coefficient that measures the response of the ith genotype to varying seasons

$\square_{ij}$ = The deviation from regression of the Ith genotype at jth season

I_j = Seasonal index, as the deviation of the mean of all genotype in jth season from grand mean such that

$$\sum_{j=1}^{n} I_j = 0$$

(a) The regression coefficient for each genotype is estimated as follows:

$$b_i = \sum_{j=1}^{n} Y_{iJ}\, I_j \,/ \sum_{j=1}^{n} I_j^2$$

(b) The mean squared deviation from linear regression (sd^2) for each genotype is estimated a follows

$$\bar{sd}^{-2} = [\sum_j d_{ij}^{\,2}/(n-2)] - (s_e^2/r)$$

Where

$$\square d_{ij}^{\,2} = [\square Y^2_{ij} / Y^2_{i.}/t] - \frac{(\Sigma Y_j\, I_j)^2}{\Sigma_j I_j^2}$$

$s_e^{\,2}$ = pooled error mean square

n = number of seasons

r = number of replications

t = number of genotypes

$I_j = (\tilde{Y}_{i\cdot} - \tilde{Y}..)$

Test of significance

a. In order to test the significance of the differences among the genotypes means i.e. $H_0 = \square_1 = \square_2 = \square_3 = \ldots = n$' the appropriate 'F' test is defined as

   $F = MS_1/MS_3$

b. To test that the genotypes do not differ for their regression on the seasonal index, i.e. $H_0 = b_1 = b_2 = \ldots = b_n$'

   $F = MS_2/MS_3$

c. individual deviation from linear regression was tested as follows

$$F = \frac{(\Sigma d_{ij}^{2})/(n-2)}{s_{e}^{2}/r}$$

14

Stability Analysis in Rice [*Oryza sativa* (L.)]

Thirugnanakumar, S. and R. Balamurugan

Stability and adaptability studies have been carried out for upland or hill rice varieties also. Amirthadevarathinam (1987) observed significant genotype × environment and reported that both linear and non-linear components were equally important for assessing stability of yield and four yield components in four environments. Both TKM 6 and Poongar were high yielding with adaptability and stability in addition to being especially suitable for unfavourable environments. Kuruvaikalangium was the best performer for all characters except grains per panicle for which PM 1023 was the best.

Dineshkumar (1987) evaluated sixteen early maturing genotypes for their stability in four major agroclimatic zones of the Karnataka state. No genotype was found stable as they observed more of unpredictable variance (pooled deviation) than predictable (genotype × environment linear).

Ganesh and Soundarapandian (1987) observed that stability of number of ear bearing tillers was a major contributor for yield stability Shivaprakash (1987) reported similar results.

Studies by Sinha and Biswas (1987) on yield and yield components of twenty photo insensitive rice cultivars revealed IR-36 and IET-4094 as high yielder with unit 'bi' and least '$S^2 d_i$' and hence they were recommended to be suitable for general cultivation for all the environments under study. Similarly IET-2881, which had high mean yield with high 'bi' was found suitable for high yielding environments.

Jaggi *et al.* (1988) identified two medium duration rice varieties, IR-36 and Usha as the most adaptive and stable varieties under varied rainfall conditions at Raipur. Narendrakulkarni *et al.* (1988) studied twenty five transplanted rice varieties including two checks at Warangal during two rainy and one winter seasons of 1983-84 and observed low regression co-efficient and least deviation of regression for grain yield in MTU-6861 and WGL-44644 and suggested that those two varieties were suitable for poor environments *i.e.,* rainy season, while MYU 2400 and WGL 22245 which had high mean with unit regression co-efficient and least deviation of regression indicating their stability for different environments (rainy and winter seasons). Among other varieties IRI 13525, C 73905 and RMR 1444 were found stable and promising.

Shivana (1989) reported that selection HP 17 characterized by medium slender grains was very promising with relatively better yield in three out of four locations in which the experiments were conducted. Sardana and Sasikumar (1990) found that pijum, CR 149- 5010-228 showed good stability because it recorded high mean yield, least deviation from regression and average regression coefficient. Majumder and Borthakur (1991) reported that in rice, the genotype × environment variances were highly significant indicating that genotypes had experienced a broad spectrum of environment and the environmental indices ranged from 11.27 to 14.09.

Samonte and Hernandez (1991) studied stability of yield and yield attributing traits of irrigated lowland rice belonging to four maturity groups in ten locations. Stability analysis of yield data revealed that IR 66 in group 1, IR 32429-122-3-1-2 and IR 42068-22-3-3-1-3 in group II, BPIR-12 in group III and IR 70 and IR 28150-84-3-3-2 in group IV were the stable genotypes for all the environments.

Ali *et al.* (1992) studied the yield stability of fourteen medium grained irrigated rice cultivars, in three environments and the results indicated that regression co-efficient ranged from -0.17 to 2.06 for average yield on environmental index and the co-efficient of determination ranged from 0.02 to 0.95. The deviation from regression mean squares ranged from 0.02 to 0.8. PK 3717-9, PK 3717-12 and PK 3849-18 showed stability over three environments and also yielded higher than the pooled mean were recommended for plant breeding programmes aimed in developing high yield potential and adaptability.

Anita and Dasgupta (1992) reported that the varieties IET 10097, IET 10109 and IET 9751 showed relatively higher degree of stability in production even under adverse situations namely low light and high water stress. De *et al.* (1992) studied genotype × environment in 47 lowland ice genotypes under four different lowland situations and reported that genotype environment were predominantly linear for ear bearing tillers per hill and non-linear for grain yield. The most stable genotypes were GR 728-1-2-2, CR 673-431 and Utkal Prava which gave overall grain yield of 432.2, 404.8 and 380.8 g per square meter respectively.

Liu (1992) conducted an experiment to know the stability of yield and yield components in six rice varieties. The experimental results showed that Jinghna 101 was the top yielder and had the best adaptability for yield stability and the first priority should be given to grain number per panicle and the second 1000 grain weight.

Mishra and Agarwal (1992) evaluated twelve early, dwarf, high yielding rice cultivars in four locations and observed that RWR 19-86-212 and RP-169 were the most stable cultivars under unfavorable conditions. Basavaraj (1993) studied sixteen genotypes showed stability for yield of which three, IET 9926, IET 5899 and Puttabatta had average stability (bi = 1); IET 99226 was rated as the best among these three based on high mean yield and desirable plant height.

Kalitha *el al.* (1993) studied twenty local rice varieties along with two widely grown controls under rainfed lowland conditions for two years. They observed that Bogabordhan out yielded both the controls yielding 4.9 t ha^{-1} and was also superior for filled spikelet percentage (96 per cent) and 1000 grain weight (30 g). Lack of significance in regression of yield across the years indicated good stability in this new cultivar.

Elsi *et al.* (1994) studied stability of medium duration rice varieties for grain yield in six cropping seasons at Pattambi. The results indicated that Cv. 18770 and Cv. 18772 were the stable genotypes for grain and straw yield with bold red grains. Geetha *et al.* (1994) identified that ADT 38 and CO45 had stable yields over four environments. Roy and Panwar (1994) evaluated ninety nine genotypes for their stability with respect to grain yield and its components genotype x environment interactions were significant for all the traits under study. The partition of genotype x environment indicated that linear component was predominant than non-linear component. The genotypes PR 103 and IR 50 were highly stable and least sensitive to environmental changes.

Major Singh *et al.* (1995) evaluated fifteen genotypes of rice were grown under six environments. They identified the genotypes RCPL 3-2, RCPL 3-6, Kanchana 4,Giza 14, MW 10 and Sarassa were stable genotypes for plant height, with regression co-efficient near to unity and non significant deviation from regression.

Silva-FG-da *et al.* (1995) studied eleven irrigated rice genotypes for stability of grain yield in different environments. They observed that the genotypes EPEAL 101, EPEAL 102 and CAN 4899 had higher stability for grain yield while, CAN 4893, CAN 3916 and METIC had poor stability for grain yield. Singh *et al.* (1995) studied promising genotypes in rice and observed significant genotype × environmental interactions. They identified genotypes stable for plant height and grain yield.

Vijendra *et al.* (1995) conducted an experiment to identify phenotypic stability of fifteen rice genotypes in four different environments. The genotype ACK 85 which was a natural mutant from IR 50 could be recommended for favourable environments in view of its above average stability for plant height, productive tillers and grain yield. Agarwal *et al.* (1996) reported that crop simulation models might facilitate the use of physiological understanding in interpreting genotype x environment interaction.

Das and Debchoudhury (1996) evaluated thirteen autumn rice genotypes in three environments under rainfed and direct seeded conditions. They observed that both linear and non-linear components of genotype x environment were significant for productive tillers per square meter and grains per panicle while, only linear component was significant for grain yield. Under medium yielding environments China, Tulasi, Annada and culture 1 were stable for grain yield while IET 10898, Tulasi, Annada and CR 633-49 for productive tillers per square meter and China and Annada for grains per panicle. Rangadoria a local genotype was found stable for grain yield under low yielding environment.

Mahapatra *et al.* (1996) studied twelve early maturing genotypes for their yield performance in thirty environments. The results revealed highly significant genotype environment interaction indicating differential response of the genotype to environmental changes. The stability analysis showed significance of linear component of variation for grain yield. Annapurna, Parijat and Keshari were stable genotypes for all environments, while Sarathi and IR 36 to high yielding environments and three tall varieties to low yielding environments.

Majumder and Borthakur (1996) studied genotype- environment interactions of some cold tolerant local and improved lines for yield and yield components. The results indicated that culture 2-1-2 and JK 2-4 (improved lines) and traditional varieties Abor B and Dullo 10 were stable in respect of yield while, PK 12-22 (improved line) and Ryllo Red (local cultivar) had average stability. Kulki another local variety and PK 1-1-3 (improved line) were unstable under cold weather conditions.

Rao *et al.* (1996) evaluated thirteen rice genotypes for their stability to grain yield and its component traits. They observed that the genotypes Aditya and Tellahamsa had regression coefficient values close to unity, indicating their highly stable performance over the test years.

Reddy and Kumar (1996) tested fourteen genotypes of rice for stability parameter over four environments. They reported that genotypes BPT 6873 and BPT 6853 had the highest stability. In a study of phenotypic stability of sixteen genotypes in rainfed rice, Sudheer *et al.* (1996) reported that there was a significant difference among the genotypes and environments. Ajmer *et al.* (1997) evaluated thirty diverse rice genotypes for stability over four environments. The results indicated highly significant genotype × environment for all the traits except days to 50 per cent flowering. IR-8, PR 106 and IT 64 were stable genotypes for grain yield.

Chikkalingaiah *et al.* (1997) labelled Ambrose as stable variety for earliness. In their study, genotype MLF-5 was found suitable for favourable environment for grain yield per hectare. The genotype MLF-18 was regarded as highly stable for grain yield per hectare. In general, no single genotype was found stable for all the characters. Sawant *et al.* (1997) studied stability of twelve early rice cultivars grown in seven locations in Maharastra and observed highly significant genotype × environment interaction. Ratnagiri 500-5-1 was found suitable for rich environments where as Basmathi 370 and Basmathi P-1 were suitable for poor environment.

Shanthakumar *et al.* (1997) found that none of the genotypes were stable for all characters. Ashvani and Dhaka (1998) evaluated thirty two promising rice genotypes grown in sixteen environments. They found that the genotypes VPR 1125-12-2-2, MDR 637, MDR 3006, HUR 51, MDR 6011, IR 8, PNR 519 and IET 1137 were the most stable for grain yield. Stability analysis of five low land rice genotypes in five low land environments showed significant genotype × environment interactions for all the traits except grain length. The effect due to the environment (linear) was significant for all characters except grain breadth and L/B ratio. Variance due to genotype × environment (linear) was significant for plant height, panicle exertion and grain yield (Basavaraj *et al.*, 1998).

Girishkumar (1998) did stability analysis of blast resistant isogenic lines. Isogenic lines (101A51) (Pi-2) and C101 LAC + C101 A51 (pi-1 + pi-2) were stable for grain yield and these lines were also found stable for traits such as plant height, productive tillers and unproductive tillers. Hedge and Vidyachandra (1998) assessed stability of rice hybrids. None of the five hybrids and four check varieties evaluated showed stability for yield over the environments. Honarnejad *et al.* (1998) estimated of stability and adaptability of eleven rice cultivars in different environmental conditions. There was no clear linear relationship between yields of varieties and environmental index.

Mahapatra and Das (1998) evaluated twelve rice varieties under three environments for their adaptation. The results indicated that among the genotypes with high mean and b_i value close to unity, it was Annapurna followed by Sarathi which had relatively lower deviation of regression were considered to have better adaptation. Mishra and Mahapatra (1998) studied yield stability of sixteen early maturing rice in various environments. The study indicated that the with high productivity were not stable over various environments.

Narendrakulkarni and Gangaram (1988) studied phenotypic stability of fifty mutant lines of Samba Mashuri for yield components and observed that among the promising mutants, the genotype × environment was significant for flowering and grain yield. The mutant lines M 7, M 14 and M 34 were found to be stable of which M 7 had good cooking quality. Prasad (1998) studied 78 doubled haploids of IR 64/ Azucena over three locations and reported twelve genotypes as stable for yield. The genotype P 138 was rated as the best among the lines.

Reddy *et al.* (1998) studied genotype × environment in twenty four low land rice genotypes under five different environments and observed significant genotype and environment interaction. Two genotypes RAU 79-214 and RAU 617-59-14-1 were found to be the most stable with high grain yield over different environments.

Hedge and Vidyachandra (1998) from their studies on five hybrids along with four standard check varieties in rice reported that IR 580 25 and AIKMR-3 with medium duration and IR 580 25 AIIR 9761 with short duration were stable for grain yield in various locations. They found that regression coefficient was near to one for all the genotypes with respect to yield components. Kandhola and Panwar (1999) conducted stability analysis of fifty two genotypes of rice for quality characters. The variation due to genotype × environment was significant for all the characters studied namely milling percentage, grain length, grain breadth and amylose content.

Lohithaswa *et al.* (1999) observed significant genotypic differences in all four seasons for all the traits except number of tillers per plant, number of productive tillers per plant and panicle length. Genotype × season interactions were found to be significant for all the seven characters considered for stability analysis. Mahapatra and Das (1999) studied twelve early maturing rice varieties for their stability over thirty environments. Genotype × environment interactions were significant for all the eight characters except 1000 grain weight. Stability was studied in twenty eight rice genotypes for late planting under tank fed areas by Suman *et al.* (1999). They concluded that no single genotype was stable for all the characters.

Abdul *et al.* (2000) reported that the regression coefficient (bi) and small deviations from regression were significant for all genotypes except PAK-4554 and Basmathi-385. Basmathi-385 performed best at all locations with average stability. Choudhury and Baruah (2000) observed that linear as well as non-linear components of genotype × environment were significant for productive tillers per hill. Shi and Yu (2000) studied genotype × environment of Indica rice for quality traits. They observed that lycine index was mainly affected by genotype × environment effects.

Honarnejad (2000) tested performance of eight Iranian rice cultivars grown at three locations. Analysis of variance showed significant differences in yield performance and location × year interaction. Ise *et al.* (2001) studied genotype × environment of high yielding cultivars in twelve locations. They observed close relationship between yield stability and some agronomic characters of rice cultivars such as early maturation, panicle density and seed fertility. Shadakshari *et al.* (2001) evaluated thirty six rice genotypes in six locations and the results indicated that IR 57773, IET 13736, Puttabatta, IET 11865, KHRS 22, KHRS 28, PUB and BKB with high mean regression coefficient equal to one and deviation from regression nearer to zero were identified as high yielding stable genotypes for grain yield.

Yang-Chang *et al.* (2001) reported that the genotype variability was as high as 66 per cent followed by G × E variability (21 per cent) and environment variability (13 per cent). The cultivars Odaebyeo, Obongbyeo and Jinbubyeo were relatively stable in all the regions. Karthikeyan (2002) studied seven rice genotypes for their stability over three environments. The variation due to genotype × environment was significant for all the characters studied. The genotypes G_{41}, G_{36} and G_{29} were the most stable genotypes.

Iftekharuddaula *et al.* (2002) observed that BR 5877-21-2-3 was the most stable genotype for growth duration *viz.*, minimum regression coefficient (b_i), insignificant deviation from regression ($S^2 d_i$) and lowest magnitude of phenotypic index (p_i). The genotypes BR 5142-24-1-1-1 and BR 4839-17-5-2-3 exhibited significant and higher b_i, insignificant S^2d_i and positive P_i values indicating that they are suitable for favourable environments.

Asenjo *et al.* (2003) reported that the AMMI explained 73 per cent of interaction sum of square, while only 57.9 per cent was explained by JRA. The AMMI model was more efficient that JRA to evaluate G × E in rice under a few environments. Azer and Ibrahim (2003) observed that the genotype ×

environment was significant for most of the parameters studied. Both linear and non-linear component of G × E were significant for days to 50 per cent heading and plant height, where as only the non-linear component of variation was significant for number of panicles, fertility percentage and grain yield. The mutant line 180/6 B/17 was the most stable under different environments.

Venkatesan (2004) studied six rice genotypes for their stability over three environments. The variation due to genotype × environment was significant for all the characters studied. The genotypes G_{41}, G_{43} and G_{46} were the most stable genotypes. Sree Krishna (2005) studied 57 genotypes for their three environments. Significant genotype × environment was observed for all the characters. The genotype G_{57} showed average stability and adapted to both favourable and unfavourable environments.

Dushyantha Kumar *et al.* (2006) studied eighteen rice genotypes were evaluated in low lands during *kharif* season of year 2001 at three locations, Mudigere, Ponnampet and Sirsi in Karnataka. Significance of both the linear and non-linear components of the G × E interaction the genotypes KHRS-21, KHRS-22 and KHRS-28 were ideally adaptable and stable. They could be recommended for cultivation in this hill zone of Karnataka. Pande *et al.* (2006) evaluated for twelve high yielding varieties (HYVS) of rice were evaluated for their adaptive advantage to various dry season rice ecologies under direct seeded wetland condition for yield and its consistency. Medium late variety Pusa 44 registered the highest yield (about 10 t/ha) when seeded in mid-November. IR 64, CR 749-20-2 and Lalat, which are very popular in boro areas, performed equally well under mid-December seeding.

Nanita Devi *et al.* (2006) reported ten promising rice genotypes for four environments. The linear components of G × E interaction was significant only for three characters, *viz.,* plant height at maturity, days to 50% flowering and grain yield plot^{-1}, whereas, the non-linear component was significant for all the characters. The stability of genotypes RCM-9 and KD-2-7-6 could be considered better for grain yield. Venkatasubbaiah (2006) studied six genotypes for their three environments. Significant genotype × environment interaction was observed for all the characters except grain breadth and kernel length indicating considerable environmental variance. The genotype G5 recorded superior mean grain yield per plant.

Anandan *et al.* (2009a) observed G x E interaction influence on grain quality of 17 rice hybrids grown in six location-year environments. Results of the study revealed that grain quality was highly influenced by environmental factors and brings out the suitability of specific genotype to specific location/ season through the biplot.

Anandan *et al*. (2009b) studied G x E interaction for grain yield of 46 genotypes by Additive Main effects and Multiplicative Interaction model generated from three saline stress environments of east coastal region of Tamil Nadu, India. The results showed highly significant genotypic and G x E interaction. The G x E interaction influenced the relative ranking of the genotypes across saline stress environment condition. Among the genotypes tested, G45, G26, G27 G35 and G34 showed low interaction effect, with score nearest to zero with above average yield. While two genotypes, G36 and G19 exhibited high positive interaction effect, gave the highest mean grain yield and is better suited to favourable environment.

The mean squares for genotypes differed among themselves for all the characters of interest excepting 100 grain weight. This indicated that the genotypes well differed among themselves in all the three environments studied. The genotype × environment mean squares were significant for days to 50 per cent flowering, plant height at maturity, number of filled grains per panicle and grain yield per plant. It indicated that the genotypes had significant interaction with the changed environments. The significant genotype × environment interaction mean squares necessitated for testing the genotypes over environments. The environmental mean squares were significant for all the seven characters of the inquiry. This amply indicated that the environment chosen were different. The result is in conformity with the findings of Saini *et al.* (1976); Naidu *et al.* (1980); Dineshkumar (1987); Amirthadevarathinam (1987); Jaggi *et al.* (1988); Sardana and Sasikumar (1990); Reddy and Chaudhary (1991); De *et al.* (1992); Roy and Panwar (1994); Singh *et al.* (1995); Mahapatra *et al.* (1996); Sudheer *et al.* (1996); Sawant *et al.* (1997); Reddy *et al.* (1998); Hedge and Vidyachandra (1998); Kandhola and Panwar (1999); Karthikeyan (2002); Venkatesan (2004); Sree Krishna (2005).

Balamurugan (2007) studied stability of 25 rice genotypes grown over three environments. The study revealed that the ratio of genotypic variance with that of the variance due to genotype × environment interaction variance indicated that the variance due to genotype were more than that of variance due to genotype × environment, implying that the varieties bred and tested at one of the environment will give the similar performance in another environment (Table 94).

Table 94. Analysis of variance for pooled data over environments

Charac-ters	Mean squares				Variance		
	Environ-ments df : 2	Genotype df : 24	G × E df : 48	Error df : 144	σ^2g	σ^2ge	σ^2g. σ^2ge
X_1	298.28**	630.83**	3.81**	1.62	69.66	0.73	1 : 0.01
X_2	391.84**	433.31**	12.72**	3.97	46.73	2.91	1: 0.06
X_3	34.86**	13.81**	1.44	1.11	1.37	0.11	1: 0.08
X_4	10.88**	8.63**	1.16	0.69	0.83	0.15	1: 0.18
X_5	1184.37**	1233.10**	80.11**	33.88	128.11	15.41	1: 0.12
X_6	4.27*	0.33	1.43	2.07	-0.12	-0.21	1: 1.75
X_7	58.28**	30.01**	3.25**	2.27	2.97	0.32	1: 0.10

* Significant at 5% level
** Significant at 1% level

ANOVA for phenotypic stability

The ANOVA for phenotypic stability indicated that the mean squares due to genotype × environment (linear) was significant for all the characters studied. The ratio of non-linear: linear component revealed the preponderance of linear component for all the characters studied. The greater proportion of the linear component of G × E interaction implied that the performance of genotypes may be predicted across environments with greater precision for all the seven traits studied (Table 95).

Table 95. Analysis of variance for phenotypic stability

Characters	Mean squares					Non- lin-ear : lin-ear ratio
	Genotype (df:24)	Environ-ments linear (df:1)	G × E linear (df: 24)	Pooled deviation (df:25)	Pooled error (df:144)	
X_1	630.83**	596.42**	4.14**	3.34**	4.86	1: 144.06
X_2	433.30**	783.50**	14.12**	10.88**	11.93	1: 55.48
X_3	13.81**	69.72**	1.10	1.71	3.35	1: 63.38
X_4	8.63**	21.79**	1.11	1.16	2.07	1:19.63
X_5	1233.10**	2368.66**	59.82**	96.40**	101.66	1: 39.59
X_6	0.33**	0.085**	0.009	0.018	6.21	1: 9.44
X_7	30.01**	116.56**	3.07**	3.29**	6.81	1: 37.96

* Significant at 5% level
** Significant at 1% level
Note: G × E is tested against pooled deviation pooled deviation is tested against pooled error

Environmental mean and environmental index

The relative ranking of the three environment indicated that E_1 was best as it showed high mean for six out of seven characters studied (Table 96).

Table 96. Environmental index (I) and environmental mean ($\bar{S}$) for different characters in rice

Characters		Environments			Overall mean
		E_1	E_2	E_3	
X_1	I	-3.4845	0.0622	3.4222	0.00
	$\bar{S}$	96.65	100.20	103.56	100.13
X_2	I	3.5349	0.7423	-4.2771	0.00
	$\bar{S}$	97.32	94.53	89.51	93.79
X_3	I	1.1307	0.0947	-1.2253	0.00
	$\bar{S}$	23.99	22.95	21.63	22.86
X_4	I	0.6271	0.0619	-0.6889	0.00
	$\bar{S}$	18.76	18.20	17.45	18.14
X_5	I	7.0505	-0.3486	-6.7019	0.00
	$\bar{S}$	146.65	139.25	132.90	139.90
X_6	I	0.0399	0.0027	-0.0426	0.00
	$\bar{S}$	1.80	1.77	1.72	1.77
X_7	I	1.4287	0.1801	-1.6088	0.00
	$\bar{S}$	27.41	26.16	24.38	25.98

Phenotypic stability parameters

Days to 50% flowering

On simultaneous consideration of the mean, regression coefficient and deviation from regression, the genotypes *viz.*, G_2, G_6, G_7, G_8, G_9, G_{11}, G_{12} and G_{23} were gathered in group I. It indicated that these genotypes had average responsiveness and were highly stable over environments. The genotypes G_{12} and G_{24} were gathered in group II, suggesting that they are suitable for unfavourable environment. The genotypes *viz.*, G_1, G_{10} and G_{20} were found to be unstable (Table 97).

Table 97. Estimates of phenotypic stability parameters -days to 50 per cent flowering (days)

Genotypes	Mean	'bi'	$\bar{S}^2 d$
G_1	92.00**	1.24	4.85*
G_2	84.33**	0.86	-1.62
G_3	107.22	0.37	2.12
G_4	113.22	1.20	-0.88
G_5	118.55	1.35	-0.05
G_6	77.11**	0.67	-1. 57
G_7	94.88**	1.05	2.32
G_8	84.33**	1.44	1.41
G_9	97.00**	0.29	0.97
G_{10}	88.00**	1.19	12.55**
G_{11}	81.22**	1.21	1.18
G_{12}	75.66**	0.48**	-1.62
G_{13}	110.55	1.11	-0.87
G_{14}	103.66	1.11	6.06*
G_{15}	117.66	0.77	0.85
G_{16}	102.66	1.15	1.34
G_{17}	105.22	0.82	1.29
G_{18}	101.66	0.82	-1.49
G_{19}	125.66	1.11	2.20
G_{20}	88.11**	1.97	5.85*
G_{21}	98.22	1.35	-1.48
G_{22}	103.33	0.86*	4.70*
G_{23}	91.55**	1.30	1.16
G_{24}	115.77	0.001*	-1.54
G_{25}	125.77	1.16	5.31*
General mean	100.13	2SE = 2.76	Mean –2SE = 97.37

* Significant at 5% level

** Significant at 1% level

Plant height at maturity

Grouping of the genotypes based on mean, regression coefficient and deviation from regression resulted in the gathering of genotypes namely, G_7, G_8, G_{12} and G_{20} had average responsiveness but highly stable over environments for this character, whereas the genotypes *viz.*, G_1, G_4, G_6, G_{14} and G_{15} were adjudged as unstable (Table 98).

Table 98. Estimates of phenotypic stability parameters - plant height at maturity (cm)

Genotypes	Mean	'bi'	$\bar{S}^2 d$
G_1	88.82**	0.88	13.90**
G_2	94.35	1.34	-2.98
G_3	97.64	0.87	-3.26
G_4	89.03**	1.00	4.82**
G_5	105.65	0.14	-3.66
G_6	72.75**	0.35	14.90**
G_7	89.13**	0.88	-3.93
G_8	81.72**	0.94	-0.50
G_9	96.98	1.37	-1.87
G_{10}	94.39	2.32	1.78
G_{11}	99.08	0.58	-3.42
G_{12}	64.16**	0.30	-3.78
G_{13}	120.38	2.57	24.06**
G_{14}	85.90**	0.77	-3.91
G_{15}	89.14**	1.18	9.22**
G_{16}	92.90	0.58	-3.89
G_{17}	95.62	0.79	-3.81
G_{18}	96.78	0.82	3.92*
G_{19}	106.18	1.40	-0.50
G_{20}	87.56**	0.65	-1.03
G_{21}	90.34	0.38	23.64**
G_{22}	108.39	2.73	3.89
G_{23}	89.98	0.89	-3.86
G_{24}	91.73	0.79	31.33**
G_{25}	116.07	0.36	-0.50
General mean	93.79	2SE = 3.92	Mean – 2SE = 89.87

* Significant at 5% level

** Significant at 1% level

Total number of tillers per plant

Simultaneous consideration of the three stability parameters *viz.,* mean, regression coefficient and deviation from regression, resulted in grouping of the genotypes *viz.,* G_1 and G_3 in group I. This indicated that these genotypes had average responsiveness and high stability for this character over environments. The genotype namely G_9 was found to be unstable (Table 99).

Table 99. Estimates of phenotypic stability parameters - total number of tillers per plant

Genotypes	Mean	'bi'	$\bar{S}^2 d$'
G_1	25.04	1.68	-0.88
G_2	22.64	0.21	-1.05
G_3	25.13	2.26	0.43
G_4	23.75	1.12*	-1.11
G_5	24.84	1.16	-0.65
G_6	19.53	1.66	-0.72
G_7	22.66	0.89	0.33
G_8	22.00	1.27	-0.66
G_9	25.44*	1.12	7.68**
G_{10}	22.64	0.11	-1.11
G_{11}	22.72	0.17	1.81
G_{12}	19.53	0.82	1.01
G_{13}	24.25	1.21	-0.38
G_{14}	23.33	1.98	-0.67
G_{15}	21.28	1.10	3.03
G_{16}	24.84	1.73	2.46
G_{17}	23.91	1.56	-0.80
G_{18}	23.06	0.45	-1.10
G_{19}	15.71	1.01*	-1.11
G_{20}	21.93	0.32	0.10
G_{21}	22.40	0.35	-0.91
G_{22}	23.51	-0.27	3.66
G_{23}	23.17	0.98	0.37
G_{24}	24.72	1.16	6.16**
G_{25}	23.37	0.85	-1.04
General mean	22.86	2SE = 2.14	Mean + 2SE = 25.00

* Significant at 5% level

** Significant at 1% level

Number of productive tillers per plant

The genotype namely, G_1 was grouped in group II, suggesting that this genotype is suitable for unfavourable environment, whereas the genotype G_{24} was gathered in group IV, indicating that it is unstable (Table 100).

Table 100. Estimates of phenotypic stability parameters - number of productive tillers per plant

Genotypes	Mean	'bi'	$\bar{S}^2 d$
G_1	20.11*	2.64*	-0.68
G_2	18.15	0.65	-0.68
G_3	19.77	2.42	-0.27
G_4	19.22	1.60	-0.63
G_5	19.24	0.62	-0.57
G_6	15.93	2.66	-0.61
G_7	17.77	-0.03	-0.10
G_8	17.54	1.02	-0.66
G_9	19.75	0.10	2.11
G_{10}	18.31	0.34	-0.66
G_{11}	17.75	-0.59*	-0.68
G_{12}	15.14	1.46	1.35
G_{13}	19.51	1.17	-0.23
G_{14}	18.56	3.38	-0.65
G_{15}	16.93	1.26	3.38
G_{16}	19.64	1.88	2.10
G_{17}	19.04	1.79	-0.56
G_{18}	18.48	0.20*	-0.69
G_{19}	12.55	1.50	-0.64
G_{20}	16.97	-1.16	2.05
G_{21}	17.57	-0.36	-0.47
G_{22}	18.70	-0.68	2.30
G_{23}	18.31	1.00	1.83
G_{24}	19.81*	1.55	5.39*
G_{25}	18.66	0.34	-0.67
General mean	18.14	2SE = 1.66	Mean + 2SE = 19.8

* Significant at 5% level

** Significant at 1% level

Number of filled grains per panicle

The genotypes *viz.*, G_1, G_2, G_{10}, G_{11} and G_{13} were gathered in group IV, suggesting that these genotypes were unstable over environments. None of the 25 genotypes was found to be stable for this trait (Table 101).

Table 101. Estimates of phenotypic stability parameters - number of filled grains per panicle

Genotypes	Mean	'bi'	$\bar{S}^2 d$'
G_1	152.15*	1.13	-26.80**
G_2	168.95**	2.15	132.64**
G_3	113.93	-0.18	-17.18**
G_4	131.24	0.64	40.15**
G_5	122.15	0.45	-30.37**
G_6	123.62	1.66	144.93**
G_7	130.57	-0.52	26.93**
G_8	143.66	2.29	417.67**
G_9	127.88	1.55	167.78**
G_{10}	165.35**	1.62	387.41**
G_{11}	210.76**	0.75	90.73**
G_{12}	126.64	0.51	57.58**
G_{13}	161.68**	2.43	-25.77**
G_{14}	123.33	0.81	-25.12**
G_{15}	134.33	0.75	104.76**
G_{16}	133.93	1.21	-32.47**
G_{17}	134.36	0.93	-28.97**
G_{18}	148.06	-0.78	-2.00
G_{19}	122.02	0.94	-23.94**
G_{20}	138.51	1.44	387.41**
G_{21}	130.15	1.04	-24.63
G_{22}	142.57	0.39	28.18**
G_{23}	134.46	1.58	-25.77**
G_{24}	138.95	0.75	2.56
G_{25}	130.66	1.39	104.04**
General mean	139.60	2SE = 11.48	Mean + 2SE = 151.08

* Significant at 5% level
** Significant at 1% level

100 grain weight

On simultaneous consideration of the mean, regression coefficient and deviation from regression coefficient, the genotypes *viz.,* G_6, G_7, G_{19}, G_{22}, G_{23} and G_{24} were found to be gathered in group I. This indicated that these genotypes had average responsiveness and high stability for this character over environments. The genotypes *viz.*, G_{20} and G_{21} were gathered in group II, indicating that these genotypes are suitable for unfavourable environment (Table 102).

Table 102. Estimates of phenotypic stability parameters - 100 grain weight (g)

Genotypes	Mean	'bi'	$'\bar{S}^2 d'$
G_1	1.67	-2.33	0.004
G_2	1.75	-0.34	-0.001
G_3	1.69	-0.79	0.17
G_4	1.36	2.62	-0.0004
G_5	1.82	0.97	-0.0021
G_6	1.92**	-0.76	0.038
G_7	2.16**	0.66	c0.071
G_8	1.64	0.34	0.070
G_9	1.50	3.04	0.0013
G_{10}	1.43	2.25	-0.0020
G_{11}	1.33	1.36	0.0039
G_{12}	1.67	-0.33	-0.0018
G_{13}	1.81	1.01	0.0000
G_{14}	1.80	-0.14	-0.0018
G_{15}	1.02	1.28	-0.0020
G_{16}	1.83	-0.29	0.0005
G_{17}	1.75	-0.69	0.0007
G_{18}	1.66	0.64	-0.0021
G_{19}	2.42**	4.35	0.0045
G_{20}	1.89**	2.33*	-0.0020
G_{21}	2.17**	0.64*	-0.0021
G_{22}	2.19**	1.12	-0.0021
G_{23}	2.07**	2.12	0.0450
G_{24}	2.28**	4.96	0.0124
G_{25}	1.30	0.92	-0.0002
General mean	1.77	2SE = 0.09	Mean + 2SE = 1.86

* Significant at 5% level

** Significant at 1% level

Grain yield per plant

The genotypes, *viz.*, G_{11} and G_{13} were found to be gathered in group I, on simultaneous consideration of the three stability parameters *viz.*, mean, regression coefficient and deviation from regression coefficient. These genotypes had average responsiveness and high stability. The genotype G_1 was found to be suitable for unfavourable environment, where as the genotypes *viz.*, G_{16}, G_{22} and G_{24} were found to be unstable (Table 103).

Table 103. Estimates of phenotypic stability parameters - grain yield per plant (g)

Genotypes	Mean	'bi'	$'\bar{S}^2 d'$
G_1	30.02**	2.18*	-2.22
G_2	25.88	0.10	-1.99
G_3	26.85	1.31	0.001
G_4	24.88	0.73	-2.15
G_5	26.58	1.33	-1.89
G_6	21.35	1.74*	-2.23
G_7	25.47	0.59	-1.91
G_8	23.58	-0.05	-2.21
G_9	26.52	0.41	1.18
G_{10}	26.52	0.16	-2.25
G_{11}	30.18**	1.97	-0.20
G_{12}	19.16	1.77	-1.02
G_{13}	29.49*	1.32	-1.71
G_{14}	26.78	2.71	1.87
G_{15}	21.15	0.98	3.04
G_{16}	29.13*	1.53	8.40**
G_{17}	28.28	1.44	-1.26
G_{18}	25.28	0.24	-2.07
G_{19}	20.29	1.57	1.19
G_{20}	24.69	-0.59	-2.25
G_{21}	26.47	-0.08	-1.86
G_{22}	29.36*	0.94	11.24**
G_{23}	26.75	0.26	6.13*
G_{24}	31.10**	0.86	19.94**
G_{25}	23.84	1.47	-0.15
General mean	25.98	2SE = 3.0	Mean + 2SE = 28.98

* Significant at 5% level
** Significant at 1% level

Similar results were reported by several workers (Narendrakulkarni, 1988; Sardama and Sasikumar, 1990; Reddy and Chaudhary, 1991; Samonte and Hernandez, 1991; De *et al.*, 1992; Anita and Dasgupta, 1992; Kalitha *et al.*,

1993; Geetha *et al.,* 1994; Elsi *et al.,* 1994; Vivekanandan and Subramanian, 1994; Singh *et al.,* 1995; Vijendra *et al.,* 1995; Das and Debchoudhary, 1996; Mahapatra *et al.,* 1996; Majumder and Borthakur, 1996; Ajmer *et al.,* 1997; Ashvani and Dhaka, 1998; Smit and Mahapatra, 1998; Shadakshari *et al.,* 2001; Karthikeyan, 2002; Venkatesan, 2004; Sree Krishna, 2005).

Table 104. Classification of genotypes based on mean, 'bi' and $\bar{s}^2_d$ for different characters

S. No.	Characters	Group I	Group II	Group III	Group IV
1.	Days to 50 per cent flowering	G_2, G_6, G_7, G_8, G_9, G_{11}, G_{12}, G_{23}	G_{12}, G_{24}	-	G_1, G_{10}, G_{20}
2.	Plant height at maturity	G_7, G_8, G_{12}, G_{20}	-	-	G_1, G_4, G_6, G_{14}, G_{15}
3.	Total number of tillers per plant	G_1, G_3	-	-	G_9
4.	Number of productive tillers per plant	-	G_1	-	G_{24}
5.	Number of filled grains per panicle	-	-	-	G_1, G_2, G_{10}, G_{11}, G_{13}
6.	100 grain weight	G_6, G_7, G_{19}, G_{22}, G_{23}, G_{24}	G_{20}, G_{21}	-	-
7.	Grain yield per plant	G_{11}, G_{13}	G_1	-	G_{16}, G_{22}, G_{24}

15

Stability Analysis in Sesame [*Sesamum indicum* (L.)]

Thirugnanakumar, S. and A. Anandan

Genotypes × Season interaction is of major importance to the plant breeder in developing stable high yielding varieties of crop plants. When new varieties are tested over a series of seasons, the relative ranking of the varieties for any given attribute is rarely the same at each season. This increases the difficulty of identifying superior stable genotypes. Comstock and Moll (1963) have shown statistically the effect of large genotype × environment interaction is reducing the progress from selection.

In order to minimize the genotype × season interaction, stratification of the seasons can be practiced. The implification in stratification is that suitable selections could be made for the infavourable and favourable seasons encountered within the sampling units proposed. Stratification of environments has been practicised for a long time to identify suitable selection that could be made for the narrower range of environments encountered within the sampling units proposed. However, even with this refinement of technique, the interactions of genotypes with locations in a sub-region and within environments encountered at the same location in different years, frequently remain too large (Allard and Bradshaw, 1964). Little is known about the environmental factors contributing to such interactions. Sprague (1966) suggested that even if such information are available, the possibility of materially reducing genotype × environment interaction in field experiments would remain questionable.

On line with Allard and Bradshaw's (1964) statement, the interactions of genotypes with locations in a sub-region with different seasons at the same location in different years, may remain too large. As far as sesame is concerned, practically nothing is known about the factors contributing to such genotype × season interactions. Hence, the progress in breeding a stable variety of sesame has been slow. Since sesame in India is grown essentially as rainfed crop, fluctuations in yield levels closely follow changes in the environment and seasons. Breeding for multiseasonal variety for may small regions within a state is being considered necessary. There is widespread awareness of achieving stability of production as well as improvement in yield of sesame under low and high-in-put conditions as well as for different seasons.

Generally selection is done in one season and subsequent evaluation of the selected genotypes is done in the other season. This may result in the elimination of the genotypes which are at selection disadvantage in the immediate season but may have ample selection advantages in the other season. Finlay (1963, 1971), Perkins and Jinks (1968b) and Bucio-Alanis *et al.* (1969) have demonstrated that stability index is a heritable trait in plants, which could be better exploited in sesame yield improvement programme.

A reliable method of estimating stability is one pre-requisite for effective selection for increased stability. Various procedures have been used to characterize individual varieties for behaviour in varying environmental conditions. Performing tests over a series of environments, which analysed in the conventional manner, give information on genotype × environment interaction, but given no measurement of stability of individual entires.

Plaisted and Peterson (1959) presented a method to characterize the stability of yield performance when several varieties are tested in a number of locations within an year. But Eberhart and Russell (1966) pointed out that this method would call for a large number of analyses if large number of varieties are to be tested.

Finlay and Wilkinson (1963) proposed a simple model for measuring the varietal adaptation. They suggested that the average yield of all cultivars grown at a particular site provide a measure of that environment for use in a regression analysis of yield stability. The mean yield and the regression could provide for selection of cultivars adapted to environments with differing productivities. When only a small fraction of the variation for genotype × environment interaction is due to heterogeneity among regression coefficients, characterization of cultivars by the coefficient was not effective (Baker, 1969; Shukla, 1972; Freeman, 1973).

Later, Eberhart and Russell (1966) modified and elaborated this method based on the regression technique for measuring the stability of genotypes. This method of stability analysis was then used by many workers (Baker, 1969; Breese, 1969; Johnson *et al.*, 1968; Perkins and Jinks, 1968a; Walton, 1976; Singh and Singh, 1980). Eberhart and Russell's method was preferable because of its explicit nature (Jowett, 1972).

According to Eberhart and Russell (1966) the phenotypic stability could be measured by three parameters, namely mean performance over environments, the linear regression (b) and deviation from regression function $\bar{S}d^2$.

Thirugnanakumar (1991) studied sixty genotypes of diverse eco-geographic origin consisting 20 large, 20 medium and 20 small seeded lines which were raised in three seasons *viz*., September 1988, January 1989 and June, 1989. The data were recorded on 16 quantitative traits comprising the seed size, yield and their components in all the three seasons and formed the testing for the effective evaluation of the breeding material.

Analysis of variance

Analysis of variance in individual seasons and in pooled analysis indicated that the genotypes differed among themselves for all the sixteen characters studied except the plant height at maturity in season I and in pooled analysis.

From the pooled analysis of variance, it was seen that the individual seasonal effect were highly significant for all the traits indicating differential effect of each season. The relative ranking of the three seasons in the present study revealed that SI was best for plant height, number of branches, number of capsules, seed length, seed breadth and oil content. SIII was best for days to maturity, number of flowers, 1000 seed weight, 1000 seed volume, seed yield and harvest index. SII was best for capsule volume, number of seeds per capsule, seed density and total dry matter production. SII seemed to be the stress seasons as it showed better expression for only four yield contributing traits. SIII seemed to the best season as it showed maximum expression of seed yield, 1000 seed weight and harvest index. Thus the widely different seasons of the investigation would thus make it possible for early selection of genotypes to create varieties well adapted to fluctuating seasons (Table 105).

Different genotypes reacted different to varying seasons which revealed the presence of highly significant genotype × season interaction mean squares for all the characters except plant height at maturity. Highly significant interaction of season on yield and yield components was also reported in many other crops such as soyabean by Schutz and Bernard (1967) and in tobacco by Gupton *et al*. (1974). In sesame, Kandasamy (1985) reported significant G×E interaction effects for yield, plant height, capsule length and number of seeds per capsule. This indicated that the genotypes of sesame like other crop showed differential responses when grown in different seasons, necessitating the need for conducting tests during several seasons.

The arithmetic translation of the mean squares into the estimates of genotypes component (s^2g) and genotype × season interaction component (s^2 gl) indicated the relative importance of the corresponding source of variations. Computation of s^2g; s^2g1 ratio revealed the major share of the genotype × season component rather than the genotypic component is determining the character expression for 1000 seed weight (1:3:71), seed breadth, capsule volume (1:3.33), number of seeds per capsule (1:2.97), number of branches (1:2.77), harvest index (1:1.57), total dry matter production (1:1.55), oil content (1:1.54), seed yield (1:1.47), days to maturity (1:1.45), number of flowers (1:1.43) and number of capsules (1:1.43). For seed length and seed density the share of the genotypic component was equal to that of the genotype × season component (1:1.00). For plant height (1:0.93) and 1000 seed volume (1:0.72) the share of the genotypic component was more than of genotype × season component. The prominence of s^2g1 compared to s^2g for most of the characters indicated the necessity for testing the breeding materials in a number of seasons. For the characters seed length and seed density, the equal share of the s^2g and s^2g1 indicated the equal importance of G×S component. A low G × S interaction component for characters has great significance for the plant breeder. In such a case, a variety breed and tested at one of the seasons will given a similar performance at other seasons (Table 106).

The G×S interaction effects were further partitioned into linear and non-linear components. The mean squares due to regression (linear component of G×S interaction) were significant for days to maturity, plant height at maturity, number of branches, number of flowers, number of capsules, number of seeds per capsule, 1000 seed volume, oil content, seed yield, total dry matter production and harvest index. The non-linear component (pooled deviations) of the G × S interactions were also significant for the above mentioned characters except 1000 seed volume, when tested against their respective pooled error. The ratio of non-linear: linear component revealed preponderance of linear component for all the characters except capsule volume and total dry matter production. For capsule volume the linear and non-linear component was almost equal, while for total dry matter production, linear component was less than the non-linear component. Predominant linear component suggested that the performance of the genotypes may be predicted across the seasons with great precision (Breese, 1969; Samuel *et al.*, 1970; Singh and Singh, 1980). This response is not only predictable but perhaps is also simply inherited (Bains and Gupta, 1972; Kumar, 1975).

The ratio of non-linear to linear component was the widest for 1000 seed volume (1:40.3) followed by plant height (1:21.99), days to maturity (1:12.93).

For 1000 seed weight and number of seeds per capsule the linear component was six and two times greater than the non-linear component, respectively. For seed yield also the linear component was two times higher than the non-linear component. The prediction with precision is possible for these characters. The other characters excluding capsule volume and total dry matter production the linear component was marginally higher than the non-linear component. So the prediction of the characters with precision is less possible comparatively. The prediction with precision is hardly possible for capsule volume where linear and non-linear component was more or less equal and total dry matter production where the linear component was slightly less than the non-linear component. The performance of these characters depend on seasonal influence on their expression (Table 107).

Stability of seeds, size, yield and component characters

Eberhart and Russell (1966) defined a stable genotype as one which showed a high mean yield, regression coefficient (b) around unity and deviation from regression nearer to zero. Later on Breese (1969), Samuel *et al.* (1970), Paroda and Hayes (1971), Cross (1977) and Chaudhary and Paroda (1980) advocated that linear regression (b) could simply be regard as a measure of response of a particular genotypes, whereas the deviation from regression ($\bar{S}d^2$) as a measure of stability.

Grafius (1956) emphasized that the studies of individual yield component can lead to simplification in genetic explanation of yield stability. If traits associated with high yield stability could be found, the plan breeder might effectively select for yield stability by selecting for these correlated traits. For that the genetic association of the component trait with yield should also be known.

The genotypes which showed high mean (overall mean ± 2SE) were considered for classification and characterization. These genotypes were classified into four groups according to the methodology followed by Mehra and Ramanujam (1997) and Singh and Singh (1980).

Group	Mean	b	$\bar{S}d^2$
I	High mean	Around Unity*	Around zero*
II	High mean	Significantly deviating from unity	Around zero*
III	High mean	Significantly deviating from unity	Significantly deviating from zero
IV	High mean	Around unity*	Significantly deviating from zero

*Deviation not significant

Population falling in group I will have average responsiveness and highly stable over seasons. Populations in group II will have above or below average responsiveness, that is well suited for stress or favourable environments and will be stable in the respective environments. Behaviour of the population falling in group III and IV cannot be predicted. Association analysis was also made among different characters based on the mean, 'b' and $\bar{S}d^2$.

Days to maturity

When all the three parameters of stability *viz.*, low mean performance 'b' and $\bar{S}d^2$ were considered together the genotypes could be classified as follows:

Group I	Group II	Group III	Group IV
G15, G19, G25	G4, G23,		G2, G8, G13
G29, G40, G42	G37, G51,	-	G20, G38, G43
G48	G52		G45, G46, G49, G50

Seven genotypes (group I) showed non significant G×S interaction for days to maturity. Days to maturity showed non-significant association with seed yield in the pooled analysis (Thirugnanakumar, 1991). While the stability parameters 'b' of days to maturity had strong negative association with 'b' of seed yield, $\bar{S}d^2$ of days to maturity showed non significant association with the $\bar{S}d^2$ of seed yield. So the responsiveness of days to maturity had negative effect on the responsiveness of seed yield. But at the same time, the stability of days to maturity had no impact on the stability of seed yield. In the absence of any relationship between the stability of days to maturity and seed yield, selection from among the range of genotypes, that showed low mean, may be effected for enhancing the yield potential only, provided other traits correlated with yield stability did not interfere.

Height of the plant maturity

A simultaneous consideration of the three parameters of stability for individual genotype led to the grouping of the genotypes as follows:

Group I	Group II	Group III	Group IV
G3, G4, G5, G6	G8, G11		G1, G12, G14
G21, G27, G32, G33		-	G16, G18, G20
G34, G35, G38			G26, G30, G31, G36
G39, G44, G49, G58			G52

In fifteen lines of group I, G×S interaction was absent. Genotypes G8 and G11 are suitable for stress of favourable seasons. Mean plant height was highly significant and positively associated with mean seed yield. No association was observed between 'b' and Sd^{-2} of plant height with 'b' and Sd^2 of seed yield (Thirugnanakumar, 1991). This suggested that the stability and responsiveness of plant height at maturity had no influence on the stability and responsiveness of seed yield.

Number of branches

Depending upon the type of interaction exhibited by individual genotypes and considering the stability parameters jointly, the genotypes were grouped as follows:

Group I	Group II	Group III	Group IV
G2, G3, G5, G6			G17, G18, G20, G42
G7, G11, G22, G25	G1, G29	-	
G26, G27, G30, G31, G33, G39, G52, G53, G54, G57, G58, G60			

Twenty genotypes fall in group I and so they had average responsiveness over seasons and they were also stable. Genotypes G1 and G29 were responsive to favourable or unfavourable seasons and stable. Number of branches showed highly significant positive association with seed yield. But there were no association between 'b' and $\bar{S}d^2$ of number of branches and 'b' and $\bar{S}d^2$ of seed yield. So, the stability and responsiveness of number of branches had no influence on the stability and responsiveness of seed yield.

Number of flowers

A simultaneous consideration of the three stability parameters led to the grouping of the genotypes as follows:

Group I	Group II	Group III	Group IV
G3, G4, G5, G7, G8	G14, G16, G35	-	-
G11, G12, G18, G20			
G26, G27, G30, G31			
G32, G33, G36, G37			
G40, G44, G49, G53, G58			

G×S interaction was absent for 22 genotypes which came under group I. Thus they exhibited average responsiveness and stability over seasons. Genotypes G14, G16 and G35 were suitable for stress or favourable seasons. This trait was highly significant and positively associated with seed yield (Thirugnanakumar, 1991). There was no relationship between 'b' of number of flowers and 'b' of seed yield as also between $\bar{S}d^2$ of number of flowers and $\bar{S}d^2$ of seed yield. This implies that the stability of seed yield will in no way be affected by the stability or instability of number of flowers.

Number of capsules

On consideration of the stability parameters the population was grouped as follows:

Group I	Group II	Group III	Group IV
G4, G6, G7, G12, G14, G31, G35, G36, G40, G44, G58	G3, G16, G30, G38	-	G11, G18, G20, G24, G26, G27, G32, G33

Eleven genotypes showed non-significant G × S interaction for number of capsules (Group I). Four genotypes fell in Group II were suitable for stress or favourable seasons. The other eight genotypes exhibited unpredictable behaviour of the G × S interaction (Group IV). Number of capsules showed highly significant positive association with seed yield (Thirugnanakumar, 1991). The 'b' and $\bar{S}d^2$ of this trait had non-significant association with 'b' and Sd^2 of seed yield. This implies that the stability and responsiveness of number of flowers had no influence on the stability and responsiveness of the seed yield.

Volume of capsules

An appraisal of the stability and mean volume of capsules of the individual genotypes on the basis of three stability parameters led to the grouping of the genotypes as follows.

Group I	Group II	Group III	Group IV
G1,G17,G32, G33,G44,G46,G47	G14,G45	-	G4,G6,G7,G12,G13, G18,G20,G28,G29,G30,G38, G39,G48,G49,G52, G48,G59

Seven genotypes showed non-significant G × S interaction for volume of capsules (Group I). Seventeen genotypes exhibited unpredictable behaviour of the G×S interaction and their performance was predictable across the seasons. Volume of capsules showed non-significant association with seed yield (Thirugnanakumar, 1991). There was a negative significant association between 'b' of capsule volume and 'b' of seed yield. So, the responsiveness of capsule volume had a negative impact on the responsiveness of the seed yield. But, there was non-significant association between the $\bar{S}d^2$ of capsule volume and $\bar{S}d^2$ of seed yield. Hence the stability of capsule volume had no impact on the stability of the seed yield.

Number of seeds per capsule

A simultaneous consideration of the three stability parameters led to the grouping of the genotypes as follows:

Group I	Group II	Group III	Group IV
G6, G7, G17, G24, G21,G48, G39, G43, G46, G47, G51, G52	G21, G48	-	G1, G4, G8, G12, G13, G14, G22, G29, G30, G33, G34, G38, G45, G50

Ten genotypes were classified in group I and thereby exhibited average responsiveness over seasons and they were also stable. Two genotypes *viz.*, G21, and G48 came in Group II, and were responsive to favourable or stress season and stable. The behaviour of fourteen genotypes falling in group IV were not predictable. The correlation between this trait and seed yield was non significant (Thirugnanakumar, 1991). The correlations of 'b' and $\bar{S}d^2$ of this trait with 'b' and $\bar{S}d^2$ of seed yield were also not significant. This implies that the responsiveness and stability of number of seeds per capsule had no influence with the responsiveness and stability of seed yield.

Length of seeds

Depending upon the type of interaction exhibited by individual genotypes and considering the stability parameters jointly, the genotypes were grouped as follows:

Group I	Group II	Group III	Group IV
G6, G9, G10, G13, G18, G31, G40, G60	-	-	G3, G4, G5, G7, G8, G16, G22, G28, G36, G44, G45, G48, 50

Eight genotypes (Group I) showed non significant G×S interaction for seed length and there by exhibited average responsiveness and stability. The behaviour of the thirteen genotypes falling in group IV were not predictable. The correlation between seed length and seeds yield was non significant (Thirugnankumar, 1991). Association among the 'b' s of seed length and seed yield was also non significant. Similarly, there was non-significant association between $\bar{Sd}^2$ of seed length and seed yield. So the stability and responsiveness of seed yield was not affected by the stability and responsiveness of the seed length.

Breadth of seeds

When all the three parameters of stability viz., high mean performance 'b' and $\bar{Sd}^2$ were considered together the genotypes could be classified as follows:

Group I	Group II	Group III	Group IV
G4, G9, G11, G12, G21, G24, G25, G26, G28, G29, G36, G41, G44, G45, G46, G58	G40	-	G15, G19, G20, G32, G38, G39, G43

Sixteen genotypes showed non-significant G×S interaction for breadth of seeds and thus displayed average responsiveness and high stability. Genotype G40 falling in group II was suited for favourable or stress seasons. The behaviour of the seven genotypes falling in group IV were not predictable. Seed breadth evinced non significant association with seed yield. The 'b' and $\bar{Sd}^2$ of seed breadth also evidenced non significant association with the 'b' and $\bar{Sd}^2$ seed yield. Therefore, the responsiveness and stability of seed breadth had no way affected the responsiveness and stability of seed yield.

Weight of 1000 seeds

A simultaneous consideration of the three stability parameters led to the grouping of the genotypes as follows:

Group I	Group II	Group III	Group IV
G6, G9, G12, G20, G27, G32, G38	G34	-	G3, G8, G10,G11, G15, G18, G19, G21, G26, G31, G33, G33, G40, G57, G60

In seven genotypes G×S interaction was absent and evinced average responsiveness and high stability. Only one genotype namely G34 evidenced linear G×S interaction ('b' significantly deviating from unity) and so its performance was predictable over seasons. It was categorized to be suited to favourable or stress seasons. Thousand seed weight was highly significantly and positively associated with seed yield (Thirugnanakumar, 1991). There were no relationship between 'b' of 1000 seed weight and 'b' of seed yield as also between $\bar{S}d^2$ of 1000 seed weight and $\bar{S}d^2$ of seed yield. This implies that the stability of seed yield will no way be affected by the stability or instability of 1000 seed weight. Similarly the responsiveness of seed yield will no way be affected by the responsiveness or non-responsiveness of the 1000 seed weight. Therefore, it would be reasonable to state that the seed yield potential could be greatly enhanced without disturbing its stability, by selecting families with high 1000 seed weight. Seven genotypes included in group I merit consideration for such a selection. Genotype G34 in Group II was specifically suited to either favourable or stress season. The fourteen genotypes of Group IV, which show high G × S interaction may also be considered for selection as they recorded high 1000 seed weight.

Table 105. Analysis of variance for pooled data over seasons

Characters	Means Squares				Variance		
	Seasons (DF.2)	Genotypes (DF.59)	Genotypes × Seasons (DF.118)	Error (DF.354)	s^2g	s^2g1	$s^2g : s^2g1$
X_1	2634.25**	55.27**	27.09**	0.64	9.11	13.23	1:1.45
X_2	47359.69**	2511.69	2041.05**	1828.33	113.89	106.36	1:0.93
X_3	26.84**	9.46**	8.76**	0.74	1.45	4.01	1:2.77
X_4	352.13**	1526.91**	743.28**	25.67	250.21	358.81	1:1.43
X_5	3039.06**	1743.29**	846.81**	26.20	286.18	410.31	1:1.43
X_6	0.06**	0.02**	0.01**	0.001	0.003	0.01	1:3.33
X_7	1060.75**	9.19**	9.15**	2.31	1.15	3.42	1:2.97
X_8	0.38**	0.09**	0.05**	0.003	0.02	0.02	1:1.00
X_9	1.03**	0.02**	0.02**	0.003	0.003	0.01	1:3.33
X_{10}	29.72**	0.47**	0.54**	0.03	0.07	0.26	1:3.71
X_{11}	194.44**	13.18**	3.18**	0.05	2.19	1.57	1:0.72
X_{12}	1.12**	0.06**	0.03**	0.002	0.01	0.01	1:1.00
X_{13}	62.56**	25.35**	13.17**	0.30	4.18	6.44	1:1.54
X_{14}	45.19**	17.68**	09.08**	0.78	2.82	4.15	1:1.47
X_{15}	33.34**	121.40**	67.12**	9.10	18.72	29.01	1:1.55
X_{16}	388.99**	186.50**	101.60**	8.01	29.75	46.80	1:1.57

* Significant at P = 5% level; ** Significant at P = 1% level; (*Source*: Thirugnanakumar, 1991)

Table 106. Analysis of variance for phenotypic stability

Characters	Mean Squares					
	Genotypes (DF 59)	Seasons (linear) (DF 1)	Genotypes × Season (linear) (DF 59)	Pooled deviation (DF 60)	Pooled error (DF 354)	Nonlinear : linear ratio
X_1	27.6356**	2633.583**	35.2734**	2.7280**	0.6370	1 : 12.93
X_2	1255.844**	47359.51**	1398.183**	63.5895**	1828.327	1 : 21.99
X_3	4.7278**	26.8453**	4.5328**	3.7366**	0.7438	1 : 01.21
X_4	763.4523**	352.5123**	368.3774	257.7895**	25.6709	1 : 01.43
X_5	871.6451**	3038.928**	441.6755**	318.444**	26.1983	1 : 01.39
X_6	0.0105	0.055	0.0069	0.0068	0.0006	1 : 01.02
X_7	4.5969**	1060.997**	13.3383**	5.8055**	2.3083	1 : 02.30
X_8	0.0445	0.3754	0.0298	0.0217	0.003	1 : 01.37
X_9	0.0099	1.0317	0.0202	0.0146	0.0033	1 : 01.38
X_{10}	0.2325	29.7199**	0.5112	0.0922	0.0278	1 : 05.54
X_{11}	6.5902**	94.4338**	3.1874**	0.0791	0.0450	1 : 40.30
X_{12}	0.0296	1.1234	0.0228	0.0154	0.0015	1 : 01.48
X_{13}	12.6764**	62.6775**	6.9974**	5.4838**	0.2965	1 : 01.28
X_{14}	8.8389**	45.1927**	4.8430**	2.5473**	0.7755	1 : 01.90
X_{15}	60.6978**	33.3129**	33.2765**	41.2487**	9.1042	1 : 00.81
X_{16}	93.2499**	388.9635**	53.1940**	46.8050**	8.0107	1 : 01.14

Note: Significance of Genotypes and Genotypes × season interaction is against pooled deviation
Significance of pooled deviation is against pooled error.

Table 107. Seasonal index (I) and seasonal mean ($\bar{s}$) for 16 characters in sesame

Characters		Seasons			
		S I	S II	S III	Overall mean
X_1	I	3.52	1.80	-5.32	000.00
	($\bar{S}$)	106.42	104.71	97.59	102.91
X_2	I	22.99	-10.31	-12.68	000.00
	($\bar{S}$)	85.57	52.36	50.07	62.67
X_3	I	0.47	-0.47	00.00	000.00
	($\bar{S}$)	9.36	8.42	8.88	8.89
X_4	I	-1.59	-0.23	1.82	000.00
	($\bar{S}$)	73.96	75.33	77.37	75.55
X_5	I	5.58	-1.39	-4.16	000.00
	($\bar{S}$)	62.53	55.56	52.76	56.95
X_6	I	-0.02	0.03	-0.01	000.00
	($\bar{S}$)	0.33	0.37	0.33	0.34
X_7	I	1.60	1.83	-3.43	000.00
	($\bar{S}$)	55.13	55.35	50.09	53.52
X_8	I	0.04	0.02	-0.06	000.00
	($\bar{S}$)	3.10	3.09	3.00	3.06
X_9	I	0.11	-0.06	-0.05	000.00
	($\bar{S}$)	1.80	1.63	1.64	1.89
X_{10}	I	-0.38	-0.18	0.56	000.00
	($\bar{S}$)	2.89	3.09	3.83	3.27
X_{11}	I	0.61	-1.46	0.85	000.00
	($\bar{S}$)	6.84	4.77	7.08	6.23
X_{12}	I	-0.11	0.09	0.02	000.00
	($\bar{S}$)	0.46	0.65	0.58	0.56
X_{13}	I	0.80	-0.20	-0.60	000.00
	($\bar{S}$)	43.35	42.35	41.95	42.55
X_{14}	I	-0.61	0.00	0.61	000.00
	($\bar{S}$)	4.38	4.99	5.60	4.99

X_{15}	I	-0.53	0.52	0.01	000.00
	($\bar{S}$)	20.39	21.44	20.94	20.92
X_{16}	I	-1.93	0.31	1.62	000.00
	($\bar{S}$)	21.79	24.04	25.35	23.73

16

Stability Analysis in Blackgram [*Vigna mungo* (L.)] Hepper

Thangavelu, P. and N. Senthilkumar

In developing countries, the most important challenges is to increase food productivity for the continuously growing population from limited land area. To great extent, this challenges has been met with good amount of success in cereals but not in case of pulses. Pulses from an indispensable component of Indian agricultural too. Further, it is an excellent source of protein to the growing population who are mostly vegetarians. The high level of lysine in the protein makes them ideal supplement to cereals. Moreover, they are endowed with nitrogen fixation, and thus improve the soil fertility.

India is one of the major pulse growing countries in the world with a total area of 23.31 million hectares, with a total production of 14.50 million tones (Source: FAO, 2011). The area, under pulses has decreased particularly during the last decades as a consequence of intensive diffusion of techniques in green revolution on the high yielding varieties of cereals. This is mainly due to the low yield potential of legumes even under high input conditions.

Among the various pulses, blackgram/urd bean known scientifically as *Vigna mungo* (L.) Hepper, former known as *Phaseolus mungo* (L.) belonging to the tribe Paseolea of family Papilionaceae is of immense importance. It is the most widely grown and highly esteemed grain legume. In India it occupies 2.9 million hectares with an annual production of 1.24 million tones (*Source*: FAO, 2011).

In Asia this crop is widely cultivated in India, Pakistan, and Bangladesh both in *kharif* and *rabi* seasons. In India, it is mainly grown in the states of Madhya Pradesh, Maharashtra, Uttar Pradesh, Rajasthan, Karnataka and Bihar. Tamilnadu is placed eleventh and tenth in area and production respectively (FAO, 2011).

The objective of blackgram improvement programme is to develop stable high yielding cultivars through systematic breeding programmes.

With a view to assess the consistency of performance of selected genotypes for stability analysis, the trial was conducted (Channa Ghouse Peera, 2011) with thirty five genotypes namely G1, G2, G3, G4, G5, G6, G7, G8, G9, G10, G11, G12, G13, G14, G15, G16, G17, G18, G19, G20, G21, G22, G23, G24, G25, G26, G27, G28, G29, G30, G31, G32, G33, G34 and G35 in a randomized block design, replicated thrice in three environments namely E1 (January 2011); E2 (August 2011) and E3 (January 2012) for stability analysis.

Since data for three environments were available from trials, it was felt that it may help to assess the stability status on the performance of genotypes by applying few important stability models. Various stability parameters were worked out on the available data.

Several authors have suggested different types of model to identify the stable genotypes. The three models commonly preferred by breeders are Eberhart and Russel (1966) model, genotype grouping technique (Francis and Kannenberg, 1978) and stability factor method (Lewis, 1954) and these three models are used in the present investigation.

1. Analysis of variance

2. G × E interaction

3. Stability parameters

The analysis of variance for individual environment as well as pooled analysis revealed significant differences among the genotypes for all the characters. Similar results were reported by several workers Singh *et al.* (1994), Manivannan (1999).

Pooled analysis showed differences among the genotypes for all the characters (Table 108). Highly significant genotype × environment interaction was observed for almost all the characters indicating that all the genotypes interacted considerably well with the environmental conditions. The results indicated the differential reactions of genotypes with different environments. The review of earlier work also indicated the presence of considerable genotype × environment interaction in blackgram, Misra (1990), Natarajan *et al.* (2001), Raje *et al.* (2001), Abbas *et al.* (2008).

After confirmation of significant genotype × environment interaction effect, it was further divided using Eberhart and Russel (1966) model. The analysis of variance for stability revealed significant differences among the genotypes for all the characters (Table 109).

Environmental (linear) variance was highly significant for all the characters except days to first flowering, number of seeds per pod and hundred seed weight. This indicated considerable additive environmental variance for all the characters except days to first flowering, number of seeds per pod and hundred seed weight. Further, variance due to genotype × environment interaction were significant for plant height, number of branches per plant, number of clusters per plant, number of pods per plant and seed yield per plant suggesting the response of genotypes, changes according to different environments. Pooled deviations were also highly significant for all the characters studied, except days to first flowering, number of pods per plant and hundred seed weight which indicated that unpredictable portion formed the major part of the genotypes×environment interaction. Similar results were obtained by Singh *et al.* (1994), Raje *et al.* (2001), Manivannan (2003), Raje *et al.* (2004).

Genotype×environment interaction (linear) was found to be significant for all the characters, except days to first flowering, number of seeds per pod and hundred grain weight indicating the contribution of linear portion to genotype×environment interaction. The more pronounced linearity of these characters indicated that variations among the genotypes could largely be explained by difference in regression slopes for these characters. The predominance of linear component noticed would help in predicting the performance of the genotypes across environment. Similar results were obtained by Natarajan *et al.* (2001), Manivannan (2003), Raje *et al.* (2004) and Bhagirath Ram *et al.* (2009), observed non-significant pooled deviations. Such contradictions arise because different workers had used different genotypes in varying environments. Thus, it can be inferred that there is no simple relationship existing between the genotypes and the environments. Therefore, it is difficult to predict the performance of all genotypes across environments.

The problem of major importance to the plant breeder in developing a variety is genotypes × environment interaction. It affects the progress to be expected in a given breeding programme. In order to minimize, genotype × environment interaction stratification of environments has been suggested. However, even with these refinements of technique, the interaction of genotypes "with locations in a sub region" and with environments encountered at the "same location" in different years, frequently remain too large (Allard and Bradhsaw, 1964).

Once the genotype × environment interaction was found significant, the next step was to identify the stable genotypes, which interact least with the environments giving a near consistent performance across environments.

Many stability models have been developed to identify such stable genotypes.

Finlay and Wilkinson (1963) in their model considered the mean value of the genotypes and regression coefficient. But, Eberhart and Russell (1966) improved this model further and they suggested that both linear and non-linear component for genotype × environment interaction should be considered in judging the phenotypic stability of particular genotype. So, in their model they considered three stability parameters *viz.*

1. Mean performance
2. Regression coefficient (b_i) i.e., regression of means of environmental index.
3. Deviation from regression ($\bar{S}d^2$).

A genotype is found to be stable based on non-significant deviation from regression coefficient and mean values. If regression coefficient is equal to unity, a genotype is considered to have average stability (similar performance in all the environments). If regression coefficient is more than unity the genotype is considered to have below average stability (good performance only in favourable environment). If regression co-efficient is less than unity the genotype is considered to have above average stability (good performance only in poor environments).

Thus depending on the character, a genotype with high or low mean (desirable mean depends on the character), unity regression coefficient and non-significant deviation from regression was considered as ideal widely adapted and stable genotype (Table 110-121).

Table 108. Analysis of variance for pooled over environments of thirty five blakcgram genotypes for various characters

Source	df	MSS								
		Days to first flowering	Plant height	Number of branches per plant	Number of clusters per plant	Number of pods per cluster	Number of pods per plant	Number of seeds per pod	Hundred seed weight	Seed yield per plant
Replication	2	0.09	1.7	0.01	0.13	0.01	0.10	0.01	0.01	0.05
Genotype	34	5.27**	88.97**	0.39	10.85**	0.86	86.82**	1.04	0.20	10.11**
Error	68	0.23	1.48	0.01	0.10	0.01	0.20	0.01	0.02	0.01

**Significant at 1% level

Table 109. Analysis of variance of thirty five blackgram genotypes for joint regression (stability) analysis for various characters

Source	df	MSS								
		Days to first flowering	Plant height	Number of branches per plant	Number of cluster per plant	Number of pods per cluster	Number of pods per plant	Number of seeds per pod	Hundred seed weight	Seed yield per plant
Replication with in environment	6	0.22	4.02**	4.28**	37.78**	1.28	1.00	1.59	5.65**	6.36**
Genotype (G)	34	5.28**	88.97**	0.39	22.13**	0.86	87.77**	1.04	0.20	10.11**
Env + (G×E)	70	0.67	34.29**	7.02**	18.13**	0.14	18.44**	0.28	8.42**	1.20
Environment (E)	2	0.20	250.78**	9.55**	26.12**	3.04*	224.67**	1.08	0.11	17.17**
G × E	68	0.69	27.92**	6.95**	17.89**	0.15	12.37**	0.25	0.08	0.73
Env (linear)	1	0.40	501.56**	0.19	52.24**	6.08*	449.35**	2.16	0.23	34.34**
G×E (linear)	34	0.63	40.62**	5.14**	31.14**	7.21**	16.54**	0.41	0.13	2.18**
Pooled deviation	35	0.73	14.78**	8.50**	4.50**	0.22	7.97**	9.44**	0.03	2.28**
Pooled error	204	0.76	4.63	9.27	37.78	1.07	1.70	1.46	6.84	3.13

**Significant at 1% level
*Significant at 5% level

Table 110. Regression coefficient for various characters in thirty five blackgram genotype

S. No.	Days to first flowering	Plant height	Number of branches per plant	Number of clusters per plant	Number of pods per cluster	Number of seeds per plant	Number of seeds per pod	Hundred seed weight	Seed yield per plant
1	0.11	-0.02	1.64	-0.11	3.29	0.96	0.76	9.12*	0.12
2	7.41	-0.82	0.39	0.99	0.98	1.74	0.99	13.67*	1.02
3	7.98	2.50	-2.51	4.49	4.35	1.72*	-0.46	-0.44	1.79*
4	-0.73	0.25	6.30	0.94	0.16	0.79*	2.68*	3.89	0.08
5	-11.06	0.88	1.99	0.37	13.44	-0.16	-1.08*	-0.28	-0.92
6	-5.41	0.97	0.09	2.87**	4.09*	-0.11	0.76	-0.31	-0.40
7	-15.10	0.55	-6.72	-3.64	3.99	1.77*	3.43	2.15	1.59
8	-3.38	3.23*	-0.27	-1.05	3.56	2.82*	0.11	-4.79	2.70
9	11.66	5.07	3.17	-0.61	3.85	2.29*	2.88	6.85	2.14
10	15.10*	-1.98	-0.03	2.18	5.52	0.28	0.83	-0.29	0.17
11	-7.61	2.98	2.88	-1.51	-17.76	1.70*	4.52*	-4.01	-0.06
12	-3.83	1.21	5.41	1.00	9.44	1.12	1.02	1.01	1.01
13	13.52*	1.01	-0.05	1.01	0.98	0.05	-3.20*	0.99	1.02
14	5.27	2.05	-4.92	0.02	0.833	-2.47	2.08*	-0.58	-0.12
15	10.73	-1.71	3.20	-2.16	-5.36	2.20	-0.36	-3.49	2.73*
16	-7.65	-2.24	0.01	2.57	3.02	1.16	0.45	0.18	0.11
17	-2.88	-0.09	3.89*	2.91	1.93	1.81	3.10	-0.46	1.85
18	-2.72	0.54	0.12	1.64	-2.21	0.84*	-5.26*	-0.57	1.29
19	1.14	0.62	2.74	-0.63	-0.82	1.79	1.28	0.43	2.45*
20	-0.02	-0.28	2.85	-3.99	-4.68	2.95	0.83	15.34	1.22
21	13.18	0.95*	0.99	1.56	-1.73	1.90*	0.95	-0.90	0.97
22	-2.35	1.45	-2.90	3.75	-2.52	-0.26	5.50	0.31	0.89

23	10.71	-0.39	-5.81	0.63	22.34*	0.61	0.51	1.83*	0.18
24	5.30	0.35	-2.03	1.78**	1.33	2.34	-0.31	3.03	2.24
25	-2.28	2.37*	0.90	-3.49	2.01	-0.16	2.43*	-0.96	-0.11
26	1.27	2.79*	3.58	-0.64	-0.99	-0.18	3.33*	2.69	-0.85
27	-6.36	1.66**	1.89	0.14	2.82*	1.84	-0.16	-2.14	0.56
28	-0.13	0.62	2.30	0.33	5.18*	1.06	-2.47	-0.12	-0.05
29	-5.85	2.45	2.39	0.31	-4.28	0.88*	-1.56	-0.19	0.19
30	5.23	3.73	2.72	0.76**	-9.97	0.37	-0.70	2.92*	2.84*
31	-0.26	2.08	-0.49*	0.62	1.56	0.68	2.61	-1.19	0.54
32	8.72	1.95	3.98	-1.42**	-0.42	-0.02	3.49	-0.64	0.09
33	-2.72	0.37	6.17	0.90	-2.23	-0.42	-3.07	0.80	2.54
34	-0.04	0.98	-0.47	0.99	1.01	1.02	7.85*	0.99	1.01
35	-1.81	-1.11	1.21	0.16	-1.62	0.82	3.23	1.76*	2.12

**Significant at 1% level
*Significant at 5% level

Table 111. Deviation of regression for various characters in thirty five blackgram genotypes

S. No.	Days to first flowering	Plant height	Number of branches per plant	Number of clusters per plant	Number of pods per cluster	Number of seeds per plant	Number of seeds per pod	Hundred seed weight	Seed yield per plant
1	1.19*	-1.21	0.09	-12.81**	-0.01	-0.39	0.02*	-0.10	0.06**
2	1.22*	20.75**	0.06	-12.75	0.01	2.30*	-0.01	-0.02*	-0.01
3	0.90*	1.60	0.11	-6.20	0.01*	-0.01	-0.01	-0.02**	0.05
4	0.97*	-1.51	0.03	-10.98	-0.01	-0.51	-0.0	-0.01	-0.01
5	1.45*	2.09	0.08	-6.59	0.12	1.42	-0.01	-0.02	0.06
6	0.15	2.02	0.06	-12.91*	-0.01	3.77**	0.03	-0.02	-0.01
7	1.84**	-0.95	0.13**	-6.62	0.08**	0.05	1.16**	-0.02*	0.18**
8	1.43*	2.04	0.04**	-11.55	0.41**	0.38	-0.01	-0.01	0.33**
9	0.64	15.86**	0.01**	-12.44	1.95**	-0.11	0.44**	0.09*	0.28**
10	-0.15	10.01**	-0.03	-12.55	0.17**	3.29**	0.02	-0.02*	0.04*
11	0.45	11.21**	0.01	-12.34	3.69**	-0.01	0.01*	-0.01	0.08**
12	0.85*	7.27**	0.03	-12.56	0.20**	2.00*	-0.01	0.03	0.0.1
13	-0.21*	7.83	0.11**	-4.96	0.01	-0.40	-0.01	-0.01	0.03
14	1.05*	5.35*	1.37**	-12.91	0.01	12.53**	0.01	-0.02	0.19**
15	-0.16	1.63	0.01**	-9.61	0.09**	77.98**	-0.01	0.01	0.09**
16	1.08*	5.26*	-0.01	-12.41	0.01	-0.08	-0.01	-0.01	4.62**
17	0.75*	19.30**	-0.01	-8.06	0.01	3.95	0.21**	0.01	0.74**
18	0.01	13.42**	-0.01	-10.71	0.02**	-0.50*	0.02*	-0.02*	0.39**
19	-0.18	17.47**	0.22**	-6.71	0.15**	29.44**	0.01	-0.02	0.01
20	-0.15	-1.34	0.02**	-1.54	0.01	16.99**	0.05**	0.50**	0.10**
21	-0.21	-1.41**	-0.01	-10.58	0.02**	-0.17	-0.01	-0.01	0.01
22	-0.20	0.05	-0.01	-8.83	0.02**	1.66*	0.26	-0.02	0.45**
23	-0.183	-0.51	0.01**	-11.68**	-0.01**	7.55**	-0.01**	-0.02	-0.01

24	0.76	-1.53	0.15**	-12.90	0.01**	4.66**	-0.01*	-0.01	0.09**
25	-0.19	-0.81	0.12**	-11.86**	0.23**	22.97**	-0.01	-0.01	0.01
26	-0.90	-1.27	0.02	-8.82	0.02	-0.37	-0.01	0.45	0.08
27	0.43	-1.52**	0.02**	-11.98**	-0.01**	2.93*	0.01	0.05**	0.35**
28	0.80*	-0.29	0.01**	-11.14	-0.01	3.87**	0.03**	0.04	0.35**
29	0.46	4.33	0.01	-11.45	0.03**	-0.54	0.01	-0.02	0.01
30	1.34*	13.90**	0.04**	-12.91	0.03**	0.14	0.10**	-0.02	0.01
31	0.83*	4.09	0.03**	-12.56	-0.01	19.14	0.15**	-0.02	0.15
32	-0.24	2.88	0.01**	-12.91*	0.11**	-0.28	0.43**	-0.02*	0.14**
33	0.54	-0.95	0.02**	-12.77	0.03**	1.78*	0.02*	-0.01	0.23**
34	0.05	-0.31	-0.01*	-12.91	0.09	0.39	-0.01	-0.01	-0.01
35	-0.06	0.45	0.01	-12.83*	0.01*	7.68**	0.17**	-0.02*	0.23**

**Significant at 1% level
*Significant at 5% level

Table 112. Mean performance of days to first flowering (days)

S. No.	Genotypes	E1 (days)	E2 (days)	E3 (days)	Pooled (days)
1	G1	31.53**	31.66*	30.13**	31.11**
2	G2	33.46	32.19	34.22	33.29
3	G3	35.98	34.90	34.04	34.97
4	G4	31.66	31.90	30.44**	31.33**
5	G5	33.89	34.40	36.36	35.22
6	G6	31.89*	32.63	33.10	32.54*
7	G7	31.12**	33.22	34.11	32.82
8	G8	33.74	34.09	35.54	34.45
9	G9	34.75	33.11	32.64	33.50
10	G10	35.86	33.62	34.19	34.56
11	G11	31.43**	32.48	33.06	32.32**
12	G12	32.68	33.38	31.79*	32.62*
13	G13	35.32	33.26	34.37	34.31
14	G14	33.65	33.00	31.87*	32.84
15	G15	34.22	32.64	32.96	33.27
16	G16	32.31	33.61	31.63*	32.52*
17	G17	32.04*	32.35	33.45	32.61
18	G18	32.01*	32.36	32.83	32.40
19	G19	32.07	32.96	32.42	32.82
20	G20	32.71	32.68	33.08	32.82
21	G21	35.01	33.00	34.08	34.03
22	G22	32.37	32.70	32.81	32.62*
23	G23	37.47	35.89	36.24	36.53
24	G24	35.29	34.62	33.66	34.52
25	G25	32.28	32.60	32.74	34.54
26	G26	34.58	34.34	34.94	34.62
27	G27	31.59**	32.45	33.10	32.38*
28	G28	32.91	33.06	31.74*	32.57*
29	G29	33.22	34.21	32.75	33.39
30	G30	34.06	33.12	35.07	34.08
31	G31	32.78	32.70	34.01	33.16
32	G32	36.33	35.04	35.33	35.57
33	G33	312.40**	31.92	30.61**	31.31**
34	G34	36.26	36.20	36.90	36.46
35	G35	33.16	33.49	32.82	33.16
General mean (days)		33.49	33.33	33.40	33.41
SE		0.71	0.72	0.70	0.28
CV (%)		2.57	2.63	2.55	1.45
CD (P=0.05)		1.42	1.44	1.40	0.78
CD (P= 0.01)		1.88	1.92	1.87	1.04

*Significant at 5 per cent level **Significant at 1 per cent level

Table 113. Mean performance of plant height (cm)

S. No.	Genotypes	E1 (days)	E2 (days)	E3 (days)	Pooled (days)
1	G1	28.68**	29.00**	29.49**	29.06**
2	G2	26.46**	32.40**	34.08**	30.98**
3	G3	48.46	36.07	47.26	43.93
4	G4	34.20*	32.96**	34.07**	33.74
5	G5	38.36	34.49	39.80	37.55
6	G6	39.85	35.47	41.02	38.78
7	G7	44.77	41.64	42.94	43.12
8	G8	52.32	34.88	45.09	44.09
9	G9	64.61	33.85**	40.15	46.20
10	G10	36.91	48.39	44.39	43.23
11	G11	54.63	35.41	35.79**	41.94
12	G12	53.71	44.48	40.25	46.15
13	G13	49.21	35.17	49.18	44.52
14	G14	44.27	34.52	44.89	41.23
15	G15	35.43	43.69	35.47**	38.19
16	G16	23.74**	34.47	23.42**	27.21**
17	G17	41.53	43.60	47.87	44.33
18	G18	34.82*	33.35**	39.31	35.83**
19	G19	39.10	34.31*	32.26**	35.22**
20	G20	40.35	41.97	41.36	41.23
21	G21	34.05*	28.96**	32.19**	31.73**
22	G22	39.04	31.94**	38.67	36.55*
23	G23	40.31	41.99	39.50	40.60
24	G24	32.24*	32.43**	33.84**	33.50**
25	G25	42.38	30.36**	40.12	37.62
26	G26	44.50	30.21**	41.18	38.63
27	G27	51.58	42.91	49.03	47.84
28	G28	43.47	40.63	49.09	42.73
29	G29	45.19	33.32	44.96	41.16
30	G30	48.50	30.50**	48.97	42.49
31	G31	39.85	29.90**	40.09	36.61*
32	G32	44.21	34.83	44.27	41.10
33	G33	31.32**	39.65**	31.83**	30.93**
34	G34	28.17**	32.72**	28.03**	29.64**
35	G35	30.65**	36.95	34.19**	33.93**

	E1	E2	E3	Pooled
General mean (days)	40.82	35.64	39.38	38.62
SE	2.86	0.60	0.80	0.70
CV (%)	8.60	2.08	2.49	3.15
CD (P=0.05)	5.72	1.22	1.60	1.96
CD (P= 0.01)	7.61	1.62	2.13	2.61

*Significant at 5 per cent level **Significant at 1 per cent level

Table 114. Mean performance of branches per plant

S. No.	Genotypes	E1 (days)	E2 (days)	E3 (days)	Pooled (days)
1	G1	2.96**	2.49	2.79**	2.75**
2	G2	2.84*	2.50	2.80**	2.71**
3	G3	2.14	2.70*	2.40	2.41
4	G4	2.85*	2.30	2.19	2.45
5	G5	3.10**	2.62	2.89**	2.87**
6	G6	2.00	2.11	1.99	2.03
7	G7	2.00	2.81**	2.70**	2.50
8	G8	2.49	2.24	2.52	2.41
9	G9	2.79*	2.44	2.46	2.57
10	G10	2.89**	2.90**	2.90**	2.90**
11	G11	2.80*	2.48	2.50	2.59
12	G12	2.73	2.69*	2.16	2.52
13	G13	1.91	2.32	1.91	2.05
14	G14	1.90	3.60**	2.41	2.64*
15	G15	2.49	2.19	2.16	2.28
16	G16	2.29	2.30	2.29	2.29
17	G17	2.99**	2.77**	2.58	2.78**
18	G18	2.35	2.34	2.34	2.34
19	G19	3.58**	2.86**	3.30**	3.25**
20	G20	2.19	1.85	1.89	1.97
21	G21	2.30	2.30	2.21	2.27
22	G22	2.80*	3.00**	3.10**	2.96**
23	G23	2.19	2.66	2.80**	2.55
24	G24	2.48	3.08**	2.69**	2.75**
25	G25	3.03**	3.41**	2.93**	3.12**
26	G26	2.40	2.00	2.03	2.14
27	G27	3.29**	2.97**	3.10**	3.12**
28	G28	3.20**	2.90**	2.96**	3.02**
29	G29	2.24	2.04	1.99	2.09
30	G30	2.60	2.19	2.31	2.37
31	G31	1.92	1.69	1.97	1.86
32	G32	2.31	2.19	1.90	2.13
33	G33	2.67	2.10	1.97	2.24
34	G34	2.76	2.79**	2.81**	2.79**
35	G35	3.08**	2.90**	2.96**	2.98**

	E1	E2	E3	Pooled
General mean (days)	2.59	2.53	2.48	2.53
SE	0.09	0.07	0.06	0.03
CV (%)	4.82	3.69	3.18	2.33
CD (P=0.05)	0.19	0.14	0.12	0.09
CD (P= 0.01)	0.26	0.18	0.16	0.12

*Significant at 5 per cent level **Significant at 1 per cent level

Table 115. Mean performance of number of clusters per plant

S. No.	Genotypes	E1 (days)	E2 (days)	E3 (days)	Pooled (days)
1	G1	7.67	7.38	7.18	7.41
2	G2	11.01**	11.80	11.76	11.52*
3	G3	8.22	16.58	14.13**	12.98**
4	G4	8.79	10.78	11.17	10.25
5	G5	13.56**	13.46	10.38	12.47**
6	G6	6.38	11.21	7.82	8.47
7	G7	12.72**	7.28	14.13**	11.38
8	G8	14.50**	12.38	12.36	13.08**
9	G9	13.12**	11.88	11.87	12.29**
10	G10	9.61	13.49	11.60	11.57*
11	G11	8.86	8.25	10.11	9.07
12	G12	10.56	7.67	8.89	9.04
13	G13	13.52**	12.94	9.81	12.09**
14	G14	10.43	10.49	10.47	10.46
15	G15	12.95**	9.82	14.22**	12.33**
16	G16	7.24	11.810	9.58	9.54
17	G17	9.06	14.64	13.60**	12.43**
18	G18	8.21	11.11	9.69	9.67
19	G19	8.28	7.93	11.28	9.16
20	G20	14.39**	6.65	7.70	9.58
21	G21	11.11**	13.32	9.90	11.44*
22	G22	10.19	17.12**	14.93**	14.08**
23	G23	11.74**	13.14	13.58**	12.82**
24	G24	9.78	12.84	10.92	11.18
25	G25	14.23**	8.61	13.72**	12.18**
26	G26	9.16	7.48	6.10	7.58
27	G27	8.68	9.20	10.05	9.31
28	G28	8.62	9.56	10.58	9.59
29	G29	6.79	7.68	8.59	7.68
30	G30	11.10**	12.40	11.53	11.67**
31	G31	12.80**	14.03*	13.93**	13.58**
32	G32	13.10**	10.67	12.23	12.00**
33	G33	12.10**	13.73	13.10**	12.79**
34	G34	7.00	7.00	7.00	7.00
35	G35	9.83	10.2	10.33	10.11

	E1	E2	E3	Pooled
General mean (days)	10.44	12.13	10.98	10.90
SE	0.20	0.76	0.75	0.19
CV (%)	2.37	8.85	8.39	3.02
CD (P=0.05)	0.40	1.75	1.50	0.53
CD (P= 0.01)	0.53	2.32	2.00	0.70

*Significant at 5 per cent level **Significant at 1 per cent level

Table 116. Mean performance of number of pods per cluster

S. No.	Genotypes	E1 (days)	E2 (days)	E3 (days)	Pooled (days)
1	G1	2.66	2.56	2.46	2.56
2	G2	2.06	2.26	2.26	2.20
3	G3	2.56	2.30	2.26	2.37
4	G4	2.86	2.90	2.86	2.87
5	G5	3.16*	2.46	2.27	2.63
6	G6	2.96	2.90	2.73	2.86
7	G7	3.23**	2.76	2.90	2.96
8	G8	3.50**	2.56	3.06*	3.04*
9	G9	3.01	4.86**	3.30**	3.72**
10	G10	3.23**	2.56	2.76	2.85
11	G11	3.66**	6.56**	5.36**	5.20**
12	G12	3.46**	2.70	2.76	2.97
13	G13	3.06	2.86	2.96	2.96
14	G14	2.83	3.00	2.83	2.88
15	G15	2.84	3.36**	3.26**	3.16**
16	G16	2.56	2.66	2.43	2.55
17	G17	3.16*	3.30**	3.10*	3.18**
18	G18	3.09	3.36**	3.28**	3.24**
19	G19	3.46**	2.93	3.37**	3.25**
20	G20	2.36	2.56	2.66	2.53
21	G21	2.70	2.96	2.86	2.84
22	G22	2.70	2.96	2.90	2.852
23	G23	4.26**	3.93**	3.00	3.73**
24	G24	2.96	2.60	2.80	2.78
25	G25	3.36**	2.66	3.07*	3.03*
26	G26	2.56	2.36	2.56	2.50
27	G27	3.06	3.00	2.90	2.98
28	G28	2.96	2.86	2.66	2.83
29	G29	2.36	2.26	2.56	2.40
30	G30	2.16	2.06	2.66	2.30
31	G31	3.06	3.00	2.96	3.01
32	G32	3.53**	3.06*	3.43**	3.34**
33	G33	2.06	2.36	2.26	2.23
34	G34	2.88	2.46	2.83	2.72
35	G35	2.81	2.68	2.86	2.78

	E1	E2	E3	Pooled
General mean (days)	2.95	2.93	2.89	2.92
SE	0.10	0.06	0.08	0.03
CV (%)	4.12	2.84	3.54	1.95
CD (P=0.05)	0.20	0.13	0.16	0.09
CD (P= 0.01)	0.26	0.17	0.21	0.12

*Significant at 5 per cent level **Significant at 1 per cent level

Table 117. Mean performance of number of pods per plant

S. No.	Genotypes	E1 (days)	E2 (days)	E3 (days)	Pooled (days)
1	G1	27.56	22.66	25	25.07
2	G2	44.63**	35.60**	38.73**	39.65**
3	G3	39.86**	31.06**	35.23**	35.38**
4	G4	30.00	26	28.56	28.18
5	G5	28.50	29.13*	27.03	28.22
6	G6	13.28	14.13	16.20	14.53
7	G7	32.53**	23.70	29.76**	28.66**
8	G8	43.60**	29.23**	36.33**	36.38**
9	G9	33.41**	21.90	29.36**	28.22
10	G10	28.13	26.43	25	26.52
11	G11	28.86	20.33	26.16	25.12
12	G12	35.43**	22.96	32.13**	30.17**
13	G13	22.73	22.40	22.10	22.41
14	G14	13.29	25.33	13.93	17.52
15	G15	28.90	16.66	12.83	19.46
16	G16	31.96**	26	28.60*	28.85**
17	G17	31.06	22.2	29.93**	27.73
18	G18	27.50	23.2	25.46	25.38
19	G19	26.40	18.03	29.60**	24.67
20	G20	34.73**	20.33	33.80**	29.62**
21	G21	39.30**	29.60**	34.43**	34.44**
22	G22	27.43	28.93**	29.90**	28.75**
23	G23	26.36	23.60	28.70**	26.22
24	G24	26.56	15.53	29.80**	23.96
25	G25	23.76	24	17.90	21.88
26	G26	19.03	20	19.96	19.66
27	G27	33.13**	24.06	31.60**	29.60**
28	G28	28	22.90	28.43	26.44
29	G29	28.93	26.06	27.56	27.52
30	G30	31.72*	29.93**	32	31.21**
31	G31	39**	34.96**	31.83**	35.26**
32	G32	26.66	26.86	27.40	26.97
33	G33	26.36	28.30*	25.30	26.65
34	G34	27.56	22.03	26.43	25.34
35	G35	38.83**	34.33**	33.40**	35.52**
General mean (days)		29.86	24.81	27.72	27.53
SE		0.69	1.64	0.43	0.26
CV (%)		2.85	8.09	1.91	1.65
CD (P=0.05)		1.40	3.28	0.88	0.73
CD (P= 0.01)		1.86	4.37	1.16	0.97

*Significant at 5 per cent level **Significant at 1 per cent level

Table 118. Mean performance of number of seeds per plant

S. No.	Genotypes	E1 (days)	E2 (days)	E3 (days)	Pooled (days)
1	G1	4.90	4.71	4.54	4.71
2	G2	4.93	4.71	4.82	4.82
3	G3	4.93	5.06**	5.11**	5.03**
4	G4	5.48**	4.56	4.83	4.96**
5	G5	4.73	5.10**	4.99**	4.94*
6	G6	4.39	4.09	4.30	4.26
7	G7	5.20*	3.60	5.36**	4.72
8	G8	4.69	4.63	4.70	4.67
9	G9	6.16**	4.92**	6.07**	5.72**
10	G10	4.25	3.90	4.20	4.11
11	G11	4.56	3.10	3.26	3.64
12	G12	4.38	4.36	4.40	4.38
13	G13	4.20	5.30**	4.99**	4.83
14	G14	4.53	3.80	4.06	4.13
15	G15	4.32	4.46	4.36	4.38
16	G16	6.16**	6.03**	6**	6.06**
17	G17	5.52	4.66	4.26	4.81
18	G18	4.32	6.03**	2.83	5.39**
19	G19	5.26**	4.99**	4.86	5.04**
20	G20	3.85	3.66	3.40	3.63
21	G21	4.67	4.46	4.56	4.56
22	G22	5.21*	3.56	3.30	4.02
23	G23	5.03	4.86*	4.88	4.92
24	G24	4.96	5.08**	5.03**	5.02**
25	G25	6.56**	5.76**	5.89**	6.07**
26	G26	5.71**	4.56	4.90*	5.05**
27	G27	3.76	3.83	3.70	3.77
28	G28	4.16	5.08**	4.60	4.61
29	G29	4.51	4.99**	5.03**	4.84
30	G30	4.56	4.67	5.06**	4.76
31	G31	5.18*	4.46	4.13	4.59
32	G32	6.17**	5.26**	4.63	5.35*
33	G33	4.13	5.23**	4.76	4.71
34	G34	6.56**	3.61	4.56	5.01**
35	G35	6.18**	4.92**	5.76**	5.62**
General mean (days)		4.97	4.64	4.72	4.78
SE		0.10	0.10	0.09	0.04
CV (%)		2.50	2.74	2.35	1.59
CD (P=0.05)		0.20	0.21	0.18	0.12
CD (P= 0.01)		0.26	0.27	0.24	0.16

*Significant at 5 per cent level **Significant at 1 per cent level

Table 119. Mean performance of hundred seed weight (g)

S. No.	Genotypes	E1 (days)	E2 (days)	E3 (days)	Pooled (days)
1	G1	5.32**	4.31	4.54	4.72
2	G2	5.62**	4.07	4.59	4.76
3	G3	4.98	5.03	5.00	5.00
4	G4	5.27**	4.86	4.86	5.00
5	G5	4.97	5.00	5.00	4.99
6	G6	4.66	4.70	4.70	4.68
7	G7	5.33**	5.10	5.13**	5.18*
8	G8	4.50	5.06	4.76	4.77
9	G9	5.36**	4.50	5.20**	5.02
10	G10	4.66	4.69	4.70	4.68
11	G11	5.10	5.59*	5.26**	5.32**
12	G12	4.80	5.40	4.90	5.03
13	G13	5.10	5.64*	5.36**	5.36**
14	G14	4.86	4.92	4.93	4.90
15	G15	4.46	4.81	4.90	4.723
16	G16	4.80	4.76	4.86	4.81
17	G17	4.96	4.96	5.20	5.04
18	G18	4.86	4.93	4.90	4.90
19	G19	4.83	4.79	4.76	4.79
20	G20	4.90	2.96	4.50	4.12
21	G21	4.90	4.96	5.10*	4.98
22	G22	4.90	4.86	4.86	4.87
23	G23	5.23**	5.03	5.06	5.11
24	G24	4.90	4.51	4.82	4.74
25	G25	4.63	4.71	4.80	4.71
26	G26	5.10	4.91	4.46	4.83
27	G27	4.50	4.82	4.36	4.56
28	G28	4.86	4.95	4.60	4.880
29	G29	4.86	4.88	4.90	4.88
30	G30	4.66	4.33	4.46	4.48
31	G31	4.46	4.16	4.50	4.52
32	G32	4.86	4.93	4.92	4.91
33	G33	5.56**	5.46	5.56**	5.53**
34	G34	5.16*	5.13	4.96	5.08
35	G35	5.03	4.83	4.90	4.92

	E1	E2	E3	Pooled
General mean (days)	4.94	4.83	4.86	4.88
SE	0.10	0.34	0.09	0.08
CV (%)	2.60	2.59	2.51	3.04
CD (P=0.05)	0.21	0.67	0.19	0.24
CD (P= 0.01)	0.27	0.90	0.26	0.31

*Significant at 5 per cent level **Significant at 1 per cent level

Table 120. Mean performance of hundred seed weight (g)

S. No.	Genotypes	E1 (days)	E2 (days)	E3 (days)	Pooled (days)
1	G1	9.29	9.27	9.66	9.41
2	G2	14.39**	13.93**	14.19**	14.17**
3	G3	14.17**	13.93**	14.09**	14.06**
4	G4	10.13	10.05**	10.19	10.12*
5	G5	8.94	10.29**	9.54	9.59
6	G6	6.60	7.10	6.60	6.77
7	G7	11.39**	9.08	10.40	10.29**
8	G8	12.32**	9.06	12.51**	11.30**
9	G9	11.42**	8.87	11.67**	10.65**
10	G10	10.42	10.30**	10.70**	10.47**
11	G11	9.22	9.46	9.67	9.45
12	G12	12.03**	11.62**	11.77**	11.81**
13	G13	10.37	8.61	9.76	9.58
14	G14	5.17	5.56	5.82	5.52
15	G15	12.02**	8.59	11.82**	10.81**
16	G16	9.66	8.46	6.63	8.25
17	G17	11.86**	9.00	10.21	10.36**
18	G18	8.88	6.86	7.69	7.81
19	G19	9.7	6.40	8.90	8.33
20	G20	11.08**	9.63	11.26**	10.66**
21	G21	10.83**	9.87**	10.46*	10.39**
22	G22	10.61	9.10	9.45	9.72
23	G23	8.32	8.11	8.37	8.27
24	G24	11.43	8.64	11.36**	10.48**
25	G25	9.13	9.34	9.31	9.26
26	G26	8.94	10.22**	9.58	9.58
27	G27	9.62	8.57	8.64	8.94
28	G28	5.16	5.52	6.01	5.56
29	G29	10.50	10.17**	10.25	10.31**
30	G30	12.08**	11.03**	11.98**	11.70**
31	G31	10.79**	9.87**	10.09	10.25**
32	G32	9.13	9.20	9.65	9.33
33	G33	9.48	6.37	9.57	8.47
34	G34	14.53**	13.19**	14.11**	13.94**
35	G35	11.57**	9.02	11.75**	10.78**
General mean (days)		10.32	9.27	10.10	9.90
SE		0.17	0.19	0.17	0.05
CV (%)		2.01	2.39	2.26	0.97
CD (P=0.05)		0.35	0.37	0.34	0.15
CD (P= 0.01)		0.46	0.50	0.46	0.20

*Significant at 5 per cent level **Significant at 1 per cent level

Table 121. Mean performance of thirty five blackgram genotypes for seed yield per plant (g) (S.K. Chinna Ghouse Peera P. No.57)

S. No.	Genotypes	E1 (g)	E2 (g)	E3 (g)	Pooled (g)
1	G1	9.29	9.27	9.66	9.71
2	G2	14.39**	13.93**	14.19**	14.17**
3	G3	14.17**	13.93**	14.09**	14.06
4	G4	10.13	10.05**	10.19	10.12*
5	G5	8.94	10.29**	9.54	9.59
6	G6	6.60	7.10	6.60	6.77
7	G7	11.39**	9.08	10.40	10.29**
8	G8	12.32**	9.06	12.51**	11.30**
9	G9	11.42**	8.87	11.67**	10.65**
10	G10	10.42	10.30**	10.70**	10.47**
11	G11	9.22	9.46	9.67	9.45
12	G12	12.03**	11.62**	11.77**	11.81**
13	G13	10.37	8.61	9.76	9.58
14	G14	5.17	5.56	5.82	5.52
15	G15	12.02**	8.59	11.82**	10.81**
16	G16	9.66	8.46	6.63	8.25
17	G17	11.86**	9.00	10.21	10.36**
18	G18	8.88	6.86	7.69	7.81
19	G19	9.7	6.40	8.90	8.33
20	G20	11.08**	9.63	11.26**	10.66**
21	G21	10.83**	.87**	10.46*	10.39**
22	G22	10.61*	9.10	9.45	9.72
23	G23	8.32	8.11	8.37	8.27
24	G24	11.43	8.64	11.36**	10.48**
25	G25	9.13	9.34	9.31	9.26
26	G26	8.94	10.22**	9.58	9.58
27	G27	9.62	8.57	8.64	8.94
28	G28	5.16	5.52	6.01	5.56
29	G29	10.50	10.17**	10.25	10.31**
30	G30	12.08**	11.03**	11.98**	11.70**
31	G31	10.79**	9.87**	10.09	10.25**
32	G32	9.13	9.20	9.65	9.33
33	G33	9.48	6.37	9.57	8.47
34	G34	14.53**	13.19**	14.11**	13.94**
35	G35	11.57**	9.02	11.75**	10.78**

	E1	E2	E3	Pooled
General mean (days)	10.32	9.27	10.10	9.90
SE	0.17	0.19	0.17	0.05
CV (%)	2.01	2.39	2.26	0.97
CD (P=0.05)	0.35	0.37	0.34	0.15
CD (P= 0.01)	0.46	0.50	0.46	0.20

*Significant at 5 per cent level **Significant at 1 per cent level

Days to first flowering

As per the stability parameters, genotype G1 had lowest mean for days to first flowering followed by G 33 and G4. The genotypes G1 and G4 were had low mean with significant deviation from regression and less than the unity regression coefficient value showed the above average stability of genotype, which are logically fit for unfavourable environment. The genotype G33 had low mean with non-significant deviation from regression and more than unity regression coefficient value indicating the below average stability for favourable environments. The genotypes G19, G26 showing stability for this traits as they showed non-significant deviation from regression and regression coefficient as around the unity. Similar results are accounted by Rao *et al.* (2004), Shanthi *et al.* (2007), Kuchanur *et al.* (2008).

Plant height

Genotype G16, G34 and G1 had the minimum plant height. The genotype G16 had significant deviation from regression and more than the unity for regression coefficient indicating that the performance of genotype can not be predicted. The genotype G1 had non-significant deviation from regression and less than the unity for regression coefficient indicating that the genotype found to have above average stability for unfavourable environment. The genotype G34 showed non-significant deviation from regression and regression coefficient as around the unity indicating that this genotype is stable for this trait. The genotypes G6, G13, G21 and G34 showed non-significant deviation from regression and regression coefficient as around the unity. Thus these genotypes are showing stability for this trait. The genotype G21 also showed significant pooled mean for this trait. Similar results are accounted by Roy and Parthasarathy *et al.* (2003), Swamy *et al.* (2004).

Number of branches per plant

The genotype G19, G25 and G27 had highest mean number of branches per plant. The genotypes G19 and G27 had significant deviation from regression coefficient more than the unity indicating that the performance of genotype can not be predicted. The genotype G25 had significant deviation from regression and less than the unity for regression coefficient indicating that the performance of genotype can not be predicted. Genotype G21 showed non-significant deviation from regression and regression coefficient around the unity and thse genotypes are showing stability for this trait. Similar results were reported by Aditya Pradap and Saisy Basandri (2006) in chickpea. Patiland Narkhede *et al.* (1989) found significant variance for genotype and environments for number of branches per plant.

Number of clusters per plant

The genotypes G8, G22 and G31 had maximum number of clusters per plant. The genotype G22 had non significant deviation from regression and more than the unity for regression coefficient indicating that the below average stability for favourable environment. The genotypes G31 and G8 had non-significant deviation from regression and less than the unity for regression coefficient value and thus found to have above average stability for unfavourable environment. The genotypes G2, G12 and G34 showed regression coefficient around the unity and also showed non-significant deviation from regression indicating that these genotypes are showing stability for this trait. The genotype G2 also showed significant pooled mean for this trait. Similar results are founded by Raje *et al.* (2004), Swamy *et al.* (2004).

Number of pods per cluster

Out of the thirty five genotypes studied, the genotype G9, G11 and G23 accounted maximum mean value for number pods per cluster. The genotype G11 had significant deviation from regression and more than the unity for regression coefficient indicating that the performance of genotype can not be predicted. The genotype G23 and G9 had significant regression coefficient and more than one indicated its responsiveness under favourable environment. The genotype G32 had significant deviation from regression and regression coefficient as less than the unity indicating that the performance of the genotype can not be predicted. The genotypes G2, G13 and G34 showed stability for this trait as they showed non-significant deviation from regression and regression coefficient as round the unity. Similar results are founded by Rao *et al.* (2004), Swamy *et al.* (2004).

Number of pods per plant

The highest number of pods per plant was accounted by G2, G3 followed by G8, G31 and G35. The genotypes G2 and G8 had showed non-significant deviation from regression and regression coefficient as more than the unity indicating that these genotypes had below average stability for favourable environment. The genotype G35 had significant deviation from regression and regression coefficient as around the unity indicating that the performance of this genotype cannot be predicted. The genotypes G1 and G34 showed non-significant deviation from regression and regression coefficient as around the unity and they showed stability for this trait. Similar results are founded by Kuchanur *et al.* (2008).

Number of seeds per pod

Out of the thirty five genotypes studied, the genotypes G16, G9 and G25 accounted maximum pooled mean value for number of seeds per pod. The genotype G16 had non-significant deviation from regression and less than the unity for regression coefficient showed above average stable performance under poor environment. The genotype G25 had significant regression coefficient and more than one indicated its responsiveness under favaourable environment. The genotype G9 had non-significant deviation from regression and regression coefficient as more the unity indicating that performance of this genotype cannot be predicted. The genotypes G2, G12 and G21 showed non-significant deviation from regression and regression coefficient as found the unity. Similar results are founded by Swamy *et al.* (2004).

Hundred seed weight

The genotypes G11, G13 and G33 recorded maximum hundred seed weight over environments. The genotypes G13 and G33 showed around unity of regression coefficient and non significant deviation from regression suggesting their suitability to unfavourable environmental condition. The genotype G11 had showed non-significant deviation from regression and regression coefficient as more the unity and thus the performance of this genotype cannot be predicted. The genotypes G12, G13 and G34 are showing stability as they had non-significant deviation from regression and regression coefficient as around the unity. The genotype G13 also showed significant pooled mean for this trait. Similar results are founded by Raje *et al.* (2001), Swamy *et al.* (2004), Kuchanur *et al.* (2008).

Seed yield per plant

Among thirty five genotypes studied, the genotypes G2, G3 and G34 recorded significant higher mean in all the three environments for this trait. Eighteen genotypes recorded significant pooled mean over environments. The genotypes G2, G3, G8, G12, G30 and G34 recorded significantly higher pooled mean for seed yield per plant. In the five genotypes, genotypes G2, G12 and G34 recorded non-significant deviation from regression and regression coefficient as around the unity. The genotypes G13 and G21 also recorded non-significant deviation from regression and regression coefficient as around the unity. In which, G13 had non-significantly lower mean and G21 had significantly higher mean for this trait. Similar results were reported by several workers Manivannan (1999), Natrajan *et al.* (20002), Roy and Parthasarathy *et al.* (2003), Raje *et al.* (2004), Swamy *et al.* (2004), Shanthi *et al.* (2007), Abbas *et al.* (2008), Bhagirath Ram *et al.* (2009), Pillai Arumugam *et al.* (2010).

Considering the overall performance, the genotypes G2, G12, G21 and G34 were founded promising with stable performance for seed yield and other yield contributing characters and may be used general cultivation and also further used for further breeding programmes.

Among the stable genotypes, the genotype G2 showed non-significant deviation from regression and regression coefficient as around unity for other characters namely, number of clusters per plant, number of pods per cluster and number of seeds per pod. The genotype G12 showed non-significant deviation from regression and regression coefficient as around unity for other characters namely, number of clusters per plant, number of seeds per pod and hundred seed weight. The genotypes G13 showed non-significant deviation from regression and regression coefficient as around the unity for other characters namely, plant height, number of clusters per plant, number of pods per cluster and hundred seed weight. The genotype G21 showed non-significant deviation from regression and regression coefficient as around unity for other characters namely, number of branches per plant and number of seeds per pod. The genotype G34 showed non-significant deviation from regression and regression coefficient as around unity for other characters namely, plant height, number of pods per cluster, number of pods per plant and hundred seed weight. Similar results were reported by several workers Natrajan *et al.* (2001), Raje *et al.* (2001), Revanappa *et al.* (2002), Manivannan (2003), Rao *et al.* (2004), Kuchanur *et al.* (2008).

17

Stability Analysis in Bhendi [*Abelmoschus esculentus* (L.) Moench]

Senthilkumar, N.

Ramya (2012) studied stability of 12 hybrids in three environments. The researcher selected the 12 hybrids based on the standard heterosis. The hybrids are: $P_2 \times P_5$, $P_1 \times P_5$, $P_3 \times P_5$, $P_2 \times P_7$, $P_2 \times P_3$ and $P_1 \times P_2$. Among the twelve hybrids, the cross combinations *viz.*, $P_2 \times P_5$, $P_2 \times P_7$ and $P_2 \times P_3$ showed non-significant reciprocal effects for most of the traits including fruit yield per plant. The direct hybrid $P_1 \times P_2$ showed less than 40 per cent standard heterosis for fruit yield per plant. It also exhibited low standard heterosis for number of fruits per plant. Eight hybrids *viz.*, $P_2 \times P_5$, $P_5 \times P_1$, $P_3 \times P_5$, $P_2 \times P_7$, $P_1 \times P_5$, $P_2 \times P_3$, $P_2 \times P_1$, $P_5 \times P_3$, based on standard heterosis (>40 per cent) and their parents *viz.*, P_1, P_2, P_3, P_5, P_7 were selected and grown in three different locations for finding their stable performance. Several authors have suggested different types of models to identify the stable genotypes. The model most commonly preferred by breeders is Eberhart and Russell (1966) model which was used in the present investigation. Various stability parameters were worked out on the available data and the results were discussed below.

Among the ten traits studied, days to first flowering, fruit weight and fruit girth showed non-significant genotype × environment interaction. Hence, the remaining seven traits were taken for further analysis. ANOVA across environment revealed highly significant mean sum of square for genotypes, indicating the presence of wide genotypic variability among the genotypes involved in the study for all the characters studied. High significant differences were observed for environments for the traits *viz.*, number of branches per plant, fruit length and fruit yield per plant, indicating the divergence among the growing environments and differential response of genotypes to the various environments. The main factors contributing to the divergence of the environments might be temperature differences during crop season, intensity and duration of sun shine hours. Therefore, the genotypes must be tested over an extensive range of environments for proper assessment of stability of genotypes (Salesh Kumar Jindal *et al.,* 2008).

The significant mean square due to genotype × environment interaction for all the characters except days to first flowering, fruit weight and fruit girth indicated that the genotypes were influenced by the environments for the expression of the characters (Ariyo and Ayovaughan, 2000; Salesh Kumar Jindal *et al.,* 2008).

Highly significant mean square due to environment (linear) indicated that the characters studied showed considerable differences under influenced environments, except number of nodes to first fruit, number of fruits per plant, plant height and fruit yield per plant. This indicated that environments were effective in influencing the performance of the genotypes.

The G×E (linear) interaction was highly significant for days to first fruit harvest, number of fruits per plant and fruit length when tested against pooled error. It indicated the differential performance of genotypes when grown over diverse environment. The two variance components of G×E interaction *viz.,* G × E (linear) and pooled deviation which were highly significant for the traits days to first fruit harvest and number of fruits per plant revealed that interactions were equally contributed by linear and non-linear component (pooled deviation) for these traits. It also suggested that predictable as well as unpredictable components contributed towards the differences in stability of genotypes, whereas, in case of fruit length, linear component regulated the G×E interaction as it was evident from its significant genotype × environment (linear) against pooled deviation. It suggested that the prediction of performance across the environments against these characters would be highly effective.

The non-linear component alone were significant for number of nodes to first fruit, days to first fruit harvest, number of branches per plant, number of fruits per plant, plant height and fruit yield per plant which elucidated substantial differences in stability of performance among the genotypes across the environments and the major part of G×E interaction was absolutely unpredictable. The unexplainable deviations from the regression on the environmental index were accounted for by significant pooled deviations. It indicted non-linear responses and included that part of G × E interactions that was unexplainable by additive environmental effects. Therefore, the regression co-efficient might not be adequate enough to predict the linear sensitivity of the genotypes for varying environments. These results are in accordance with earlier findings of Poshiya and Vashi (1997) and Salesh Kumar Jindal *et al.* (2008).

Number of methods are available for assessing the stability performance of different crop varieties. Lewis (1955) measured the phenotypic stability interms of stability factors based on mean performance in the highest yielding environments with that in the lowest yielding environments. Finlay and Wilkinson (1963) used linear regression as a quantitative measure of phenotypic stability to describe varietal adaptability over a range of environments. The absolute phenotypic stability of a genotype was expressed as b = 1 and an ideally adapted genotype was said to be the one having the maximum yield potential in the most favorable environment and also the maximum phenotypic stability. Eberhart and Russell (1966) suggested that both linear (bi) and non-linear ($\bar{S}^2di$) components to the genotype × environment interaction should be considered in judging the phenotypic stability of a particular genotype. Thus, a genotype with high mean $(\bar{X})$, unit regression co-efficient (bi = 1) and least square deviation from regression with non-significance ($\bar{S}^2di = 0$) is considered as an ideal, widely adapted and stable genotype. Paroda and Hayes (1971) and Jatasra and Paroda (1980) emphasized that linear regression could simply be regarded as a measure of response of a particular genotype, whereas the deviation around the regression ($\bar{S}^2di$) is the most suitable measure of stability and genotypes with least deviation around the regression line are the most stable and vice versa.

A simultaneous consideration of the three parameters *viz.,* mean performance, bi and $\bar{S}^2di$ are essential to select stable genotypes.

The hybrids namely $P_3 \times P_5$, $P_5 \times P_3$ and parent P_3 were considered as the best adapted for all the environments for the trait days to first trait harvest. The hybrids $P_2 \times P_5$, $P_5 \times P_1$ and parent P_5 had high mean value for number of branches per plant and were highly stable as indicated by unit regression co-efficient and deviation from regression. The hybrid $P_2 \times P_1$ was adapted for all the environments for the traits number of fruits per plant and plant height. For fruit length, the hybrids $P_5 \times P_1$ and $P_3 \times P_5$ were stable and for fruit yield per plant, the hybrids $P_5 \times P_1$, $P_3 \times P_5$ and the parent P_5 were stable and suitable for all the environments.

The parent P_2 possessed higher mean value and regression co-efficient greater than unity with non-significant $\bar{S}^2di$ value for the trait number of nodes to first fruit and days to first fruit harvest. The parent P_2 was suitable for favourable environment for above traits. The hybrid $P_2 \times P_3$ and parent P_1 were suitable for high input/favourable environment for number of fruits per plant and for fruit length, the hybrid $P_2 \times P_5$ was best suited for favourable environment.

Regarding the days to first fruit harvest, the hybrids $P_2 \times P_5$, $P_2 \times P_7$ and $P_2 \times P_3$ had high mean value with non-significant $\bar{S}^2di$ and negative regression co-efficient and thus were found to be suitable for poor/unfavourable environment. The hybrid $P_3 \times P_5$ was found to be suited for unfavourable environment for number of branches per plant and the parent P_3 for number of fruits per plant. For the trait fruit length, hybrid $P_5 \times P_3$ and parents P_1, P_3 and P_5 were suited for unfavourable or poor environment. The parent P_1 is best suited for unfavourable environment for the trait plant height (Tables 122-124).

Table 122. Performance of stable/average responsive genotypes ($\bar{S}^2di$ – non-significant and bi – around unity)

Days to first fruit harvest	Number of branches per plant	Number of fruits per plant	Fruit length (cm)	Plant height (cm)	Fruit yield per plant (g)
$P_3 \times P_5$ (44.18)	$P_2 \times P_5$ (4.24)	$P_2 \times P_1$ (30.31)	$P_5 \times P_1$ (17.29)	$P_2 \times P_1$ (117.86)	$P_5 \times P_1$ (535.99)
$P_5 \times P_3$ (44.31)	$P_5 \times P_1$ (4.16)		$P_3 \times P_5$ (18.37)		$P_3 \times P_5$ (514.83)
P_3 (45.87)	P_5 (3.78)				P_5 (374.91)

Table 123. Genotypes suitable for high input/favourable environment ($\bar{S}^2di$ – non-significant and bi – more than unity)

Number of nodes to first fruit	Days to first fruit harvest	Number of fruits per plant	Fruit length (cm)
P_2 (2.07)	P_2 (48.49)	$P_2 \times P_5$ (32.07) P_1 (26.98)	$P_2 \times P_5$ (18.53)

Table 124. Genotypes suitable for low input/unfavourable environment($\bar{S}^2$di – non-significant and bi – less than unity)

Days to first fruit harvest	Number of branches per plant	Number of fruits per plant	Fruit length (cm)	Plant height (cm)
$P_2 \times P_5$ (44.82)	$P_3 \times P_5$ (4.16)	P_3 (24.78)	$P_5 \times P_3$ (16.74)	P_1 (97.41)
$P_2 \times P_7$ (43.83)			P_1 (16.18)	
$P_2 \times P_3$ (43.13)			P_3 (16.05)	
			P_5 (18.40)	

References

Abdul, Q., M.Y. Multi and S.A. Rabbani. 2000. Evaluation of different rice genotypes for stability in yield performance. *Pakistan J. Sci. Indust. Res.*, 43(3): 276-281.

Agarwal, P.K., M.J. Kropff, R.B. Mathews, C.G. McLaren, M. Cooper and G.L. Hanmer. 1996. Using simulation models to design a new plant type and analyse genotype by environment interactions in rice. *Plant Adapt. Crop Improv.*, 403-418.

Ajmer, S., B.S. Chabra, P.S. Sabharwal and A. Singh. 1997. Stability analysis for yield and its components in rice (*Oryza sativa* L). *Indian J. Agric. Res.*, 31(3): 149-155.

Ali, S.S., S.J.H. Jafri, R.A. Faiz, S. Mahamood and M.A. Butt. 1992. Stability analysis for irrigated rice yield. *IRRN*, 17(5): 5-6.

Allard, R.W. and A.D. Bradshaw. 1964. Implication of genotype environmental interaction in applied plant breeding. *Crop Sci.*, 4: 503-508.

Amirthadevarathinam, A. 1987. Stability analysis of some released varieties, local cultivars and promising cultures of dry and semi-dry paddy. *Madras Agric. J.*, 21(10-11): 434-439.

Anandakumar, C.R. and M. Subramanian. 1992. Genetic divergence studies in upland rice. *Oryza*, 29: 139-141.

Anandan. A, Sabesan. T, Eswaran. R, Rajiv.G, Muthalagan.N and Suresh.R. 2009a. Appraisal of Environmental Interaction on Quality Traits of Rice by Additive Main Effects and Multiplicative Interaction Analysis. *Cereal Res. Comm.* 37(1): 139-148.

Anandan.A, Eswaran. R and Prakash.M, 2011. Diversity in Rice Genotypes under Salt Affected Soil Based on Multivariate Analysis. *Pertanika J. Trop. Agric. Sci.* 34 (1): 33–40.

Anandan.A, Eswaran. R, Sabesan. T, and Prakash.M, 2009b. Additive Main Effects and Multiplicative Interactions Analysis of Yield Performances in Rice Genotypes under Coastal Saline Environments. *Adv. Biol. Res.* 3 (1-2): 43-47.

Anita, S. and D.K. Dasgupta. 1992. Screening of rice varieties for stability in production under low light and high water stresses during wet season. *Oryza*, 29: 177-178.

Anon, 1948. The wealth of India (Raw materials) Vol. I. Publications and Information Directorate, CSIR, New Delhi, pp. 125-126.

Anon, 1986. Annual Report. National Bureau of Plant Genetic Resource, New Delhi, p. 26.

Anon, 1988. Annual Report. National Bureau of Plant Genetic Resource, New Delhi, p. 26.

Anon, 1992. The Wealth of India (Raw materaisl) Vol. IA. Publications and Information Directorate, CSIR, New Delhi, p. 447.

Anon, 1996. ICAR Adhoc scheme on technology development for handling and utilization of common types of jackfruit. Final Report. Kerala Agricultural University, Vellanikkara, pp. 45-46.

Anon, 2002. www.ipgri.cgiar.org.

Anon. 1994. Global Aam's race, Times of India, New Delhi, June 18, 1994.

Anonymous. 1988. Personal communication from agricultural and processed food products export development authority (PAEDA)-6. Bhikaji Cama Place, New Delhi-110 066.

Anonymous. 2009. Agricultural statistics at a glance. Department of Agriculture and Co-operation, Ministry of Agriculture, Government of India. (www.dacnet.nic.in).

Arun, S., D.V. Yadav, A.K. Singh, Gaurav Yadav, Surinder Gulia and K.R. Gupta. 2002. Genetic divergence in aromatic rice. *National J. Pl. Imp.*, 4(2): 46-49.

Arunachalam, V. 1981. Genetic distance in plant breeding. *Indian J. Genet.*, 41: 226-236.

Arunachalam, V. and A. Bandyopadhyay. 1984. Limits to genetic divergence for occurrence of heterosis – Experimental evidence from crop plants. *Indian J. Genet.*, 44: 548-554.

Asenjo, C.A., R. Bezus and H.A. Acciaresi. 2003. Genotype environment -interactions in rice. *Cereal Res. Commun.* 31(1 2): 97-104.

Ashvani, P. and R.P.S. Dhaka. 1998. Phenotypic stability for grain yield in rice. *Oryza*, 35(3): 269-271.

Azer, S.A. and K.I.M. Ibrahim. 2003. Genotype Environment interaction and phenotypic stability for grain yield and its attributes in some promising rice mutant lines. *Ann. Agrl. Sci.*, 48: 159-172.

Babu, S.J., A. Malarmathi, P. Yogameenakshi, A. Sheeba and P. Rangasamy. 2003. Genetic divergence studies in rice. *Crop Res.*, 25(2): 280-286.

Bailey, I.H. 1949. Manual of cultivated plants. Macmillan Ltd., New York, p. 280.

Baker, R.J. 1969. Genotype environment interactions in yields of wheat. *Canadian J. Plant Sci.*, 49: 743-751.

Balamurugan, 2007. Studies on genetic diversity and stability in rice (*Oryza sativa* L.). M.Sc. (Ag.) Thesis, Annamalai University, Chidambaram.

Bansal, U.K., R.G. Saini, N.S. Rani and A. Kaur. 1999. Genetic divergence in quality rice. *Oryza*, 36(l): 20-23.

Barrau, J. 1976. Bread fruit and releatives. Evolution of crop plants (ed. Simmonds, N.W.). Longman, pp. 202-210.

Basavaraj, D.M. 1993. Stability analysis of promising genotypes of low land rice (*Oryza sativa* L) of the hill zone of Karnataka. M.Sc. (Ag.) Thesis, Univ. Agrl. Sci., Bangalore.

Basavaraj, D.M., H.C. Lohithaswa, E.Y. Shadakshari and R.S. Patil. 1998. Stability analysis of promising low land rice (*Oryza sativa* L.) genotypes of hill zone of Karnataka. *Karnataka J. Agric. Sci.*, 11(2): 309-373.

Batra, I.R. and P.D. Milner, 1976. Some Asian fermented foods and beverages in developments. In: Underkofler, L.A. (ed.), Industrial Microbiology, *American Institute of Biological Sciences*, Washington, DC, 17: 117-128.

Beinecke and Lowe (1969) A selection system for superior black walnut trees and other hardwoods, p. 27-33. In: Proceedings of the tenth southern conference on forest tree improvement. Houston, Texas.

Bendale, V.W., S.S. Atanur, S.C. Bhave, J.L. Mehta and U.B. Pethe. 2003a. Genetic divergence for yield and yield components in okra. *Orissa J. Hort.*, 31(1): 30-33.

Benjamin, H.B., H. K. Ehrig and D.A. Roth. 1951a. The use of okra as a plasma replacement. Rev. *Canadienne Biol.*, 10: 215-220.

Benjamin, H.B., H.K. Ehrig and D.A. Roth. 1951b. The use of plasma extender in animals. *Proc. American Vet. Medi. Assoc.*, pp. 193-196.

Berry, S.K. and Kalra, C.L. 1988. Chemistry and technology of jack fruit (Artocarpus heterophyllus). A review. *Indian Food Packer*, 8: 63-75.

Bhatnagar, H.C. and H. Subramanyam. 1973. Some aspects of preservation, processing and export of mango and its products. *Indian Food. Packer*, 27(4): 33-52.

Bhore, D.P., Ghunke, S.H. and Shelke, V.G. 1980. A promising type of jack fruit. *J. Maharashtra Agric. Univ.*, 5: 81-82.

Bisht, I.S., R.K. Mahajan and T.G. Kawalker, 1999. Diversity in greengram (*Vigna radiata* (L.) Wilczek) germplasm collection and its potential use in crop improvement. *Ann. Appl. Biol.*, 132.

Blackith, R.E. 1960. A synthesis of multivariate techniques to distinguish patterns of growth in grasshoppers. *Biometrics*, 16: 28-40.

Bose, J.K. and Mitra, S.K. 1990. Fruits tropical and subtropical. Naya Prakash, Calcutta, pp. 487.

Bose, R.D., 1932. Studies in Indian pulses. Urd or blackgram (*Phaseolus mungo* (L.) var. Roxburgii Prain). *Indian J. Agric. Sci.*, 2: 265-267.

Botstein D, White RL, Skolnick M, Davis RW 1980. Construction of a genetic linkage map in man using restriction fragment length polymorphisms. *Amer J Hum Genet.* 32: 314–331.

Brackmann, A., J. Streif and F. Bangerth. 1993. Relationship between reduced aroma production and lipid metabolism of apples after long term controlled atmosphere storage. *J. Amer. Soc. Hort. Sci.*, 118: 243-247.

Braun J, Bos MS 2004. The changing economics and politics of rice: Implications for food security, globalization, and environmental sustainability. In "Rice Is Life: Scientific Perspectives for the 21st Century" (K. Toriyama, K. L. Heong, and B. Hardy, Eds.), pp. 7–20. International Rice Research Institute, Los Banos, Philippines and Japan International Research Center for Agricultural Sciences, Tsukuba, Japan.

Breese, E.L. 1969. The measurement and significance of genotype environment interactions in grasses. *Heredity*, 24: 181-200.

Brennan, P.S. and D.E. Byth. 1979. Genotype environmental interaction for wheat yields and selection for widely adapted wheat genotypes. *Australian J. Agric. Res.*, 30: 221-232.

Brian, W. 1986. Some methods of analysing genotype environment interaction. *Heredity*, 56: 243-253.

Bucio, A.L. 1966. Environmental components of variability II. hererozygotes. *Heredity*, 21: 399-405.

Bucio, A.L. and J. Hill. 1966. Environmental and genotype environmental components of variability in inbred lines. *Heredity*, 21: 387-397.

Bucio, A.L., J.M. Perkins, M. Jean and J.L. Jinks. 1969. Environmental components of variability. *Heredity*, 24: 399-405.

Burkill, I.H. 1935. A dictionary of the economics products of the Malay Peninsula. Crown agents, London.

Byth, J.S., K.S. Kanwal and T.R. Sharma. 1986. Radiation chlorophyll mutations in rice. *Res. Dev. Rep.*, 3(2): 68-71.

Campbell, R.J. and Mc Naughton, R. 1994. New jack fruit cultivars under trail in South Florida. Proc. Intreamerican. *Soc. Trop. Hort.*, 38: pp. 201-205.

Chadha, K.L. 1994. Horticulture. Delayed recognition potential. Pp. 111-117. In: The Hindu Survey of Indian Agriculture.

Chandawat, B.S. and D.V.R. Rao. 1981. Effect of certain plant regulators and Waxel on ripening of sapota (*Manilkhara achras* (Millo Forsbers). In: National Symposium on Recent Advances in Fruit Development Organized by PAU, Ludhiana on 28-31, July 2002.

Chang, T.T. 1976. The origin, evolutions, cultivation, dissemination and diversification of Asian and African rice. *Euphytica*, 25: 425-441.

Chaturvedi, H.P. and D.M. Maurya. 2005. Genetic divergence analysis in rice (*Oryza sativa* L.). *Advances in Plant Sciences*, 18(1): 349-353.

Chaudhari, S.M., U.S. Wagh, U.T. Desai and G.M. Idate. 2006. Evaluation of mango (*Magnifera indica*) cultivars for quality attributes. *Indian J. Agri. Sci.*, 75(9): 504-506.

Chaudhary, M. and A.K. Sarangi. 2002. Genetic divergence in traditional aromatic 'rice accessions of Madhya Pradesh and Chattisgarh. *Crop Improv.*, 29: 146-150.

Chaudhary, R.C. and D.V. Tran. 2001. Specialty rice of the world – A prologue. In: R. Juffy, (ed.), Specialty rice of the world: Breeding, Production and Marketing. Food and

Agricultural Organization of the United Nations Rome, Oxford and IBH Publishing Co. Pvt. Ltd., New Delhi and Kolkata.

Chauhan, J.S. and V.S. Chauhan. 1994. Genetic divergence in upland rice (*Oryza sativa* L.). *Oryza,* 31(2): 103-106.

Chauhan, K.C. and Kanwar, M.S. (2001). Nature of variability and character associations for cone and seed characteristics in Pinus roxburghii Sargent plus trees. *Indian J. Genet.*, 61(2): 151-154.

Chaurasia, B.D. and U.K. Sharma, 1974. Karyological studies in Phaseolus mungo Linn. *Broteria ser Trimest Ciene. Nature*, 43: 33-34.

Chikkalingaiah, E. Gangappa, S. Ramesh and A. Mohanrao. 1997. Stability analysis of some scented rice genotypes. *Crop. Res.*, 13(2): 397-403.

Choudhary, R.C., J.S. Nanda and T.B.R. Nagendragupta. 1972. Adaptability index in rice. *Oryza*, 9(1): 13-21.

Choudhury, P.K.D. and D.K. Baruah. 2000. Stability analysis of some cold--tolerant boro rice genotypes. *Ann. Biol.*, (Ludhiana), 16(l): 45-49.

Cobley, L.S., W.M. Steele. 1976. An introduction to the botany of tropical crops. 2nd edition, Longman, London, pp. 146-148.

Corner, E.J.H. 1952. Wayside treeds of Malaya. Singapore Government Printer, Singapore, pp. 193-199.

CSIR (1956) The world of India (Raw Materials) Vol. IV, Council Sci. Ind. Res., New Delhi, p. 99.

Daniel Sundararaj, D. and G. Thulasidas, 1976. Botany of field crop. McMillan Company of India, New Delhi, pp. 137-139.

Das, G.R. and D.N. Borthakur. 1973. Genetic divergence in rice. *Indian J. Genet.*, 33: 436-443.

Das, P.K. and R.K. Debchoudhury. 1996. Phenotypic stability for grain yield and its components in rainfed autumn rice (*Oryza sativa* L). *Indian J. Genet.*, 56(2): 214-218.

De, D.N. and R. Krishnan, 1966. Studies on pachytene and stomatic chromosomes of *Phaseolus mungo L. Genetica*, 37: 581-587.

De, R.N. and A.V.S. Rao. 1987. Genetic divergence in rice under low land situation. *Rice Res., News.*, 88(23): 1-2.

De, R.N., J.N. Reddy, A. V.S. Rao, G. Ramakrishna and K.L. Pande. 1992. Stability of rice yield under, different lowland situations. *Indian J. Genet.*, 52(2): 139-143.

Deepthi Rai, and Misra, K.K. 2005. Studies on genetic divergence in Bael (Aegle marmelus Correa). *Indian J. Hort.*, 62(2): 152-154.

Dhaduk, L.K., D.R. Mehta and K.D. Patel. 2004. Genetic diversity in okra. *Orissa J. Hort.*, 32(1): 70-72.

Dhagat, N.K. and S.P. Singh, 1983. Genetic divergence in Kodo millet. *Indian J. Genet.*, 43(2): 168-172.

Dhamu, K.P., M.K. Ayyaswamy and M. Shanmugasubramanian. 1983. Genetic diversity in sesamum. *Madras Agric. J.*, 70: 495-498.

Dhankhar, S.K., B.S. Dhankhar and R.K. Yadava. 2008. Cluster analysis on advanced breeding lines for morphological characters and yield components in okra. *Indian J. Hort.*, 65(3): 289-292.

Dhawan, N.L. and J. Singh. 1961. Flint × Dent maize hybrids give increased yields. *Curr. Sci.*, 30: 233-234.

Dineshkumar, V. 1987. Stability analysis of yield and components in early maturing genotypes of rice (*Oryza sativa* L.). M.Sc. (Ag.) Thesis, Univ. Agrl. Sci., Bangalore.

Dushyantha Kumar, S.M. and Y.G. Shadakshari. 2006. Stability analysis for grain yield and yield components of rice (*Oryza sativa* L.) in low land of hill zone of Karnataka. *Indian J. Gent.*, 66(2): 141-142.

Dutta, S. 1957. Cultivation of jackfruit in Assam. *Indian J. Hort.*, 189-196.

Dwivedi, S.V., S.K. Pande and N.P. Shahi. 2003. Growing of vegetables – the need of hour. *Agrobios News.*, 1(a): 27-32.

Eberhart, S.A. and W.A. Russell. 1966. Stability parameters for comparing varieties. *Crop Sci.*, 6: 36-40.

Edwards, W.R. Jr. and J. C. Miller. 1947. Okra seed oil. *Chemurgic Digest*, 29: 31-33.

Elsi, C.R., C.A. Rosamma and P.V. Prabhakaran. 1994. Stability of new rice genotypes for grain yield. *J. Trop. Agric.*, 32(1):7-9.

Engler, A. and Prantl, K. 1988. Die Nutritrilichen Pflanzenfamilien. Leipzig, pp. 80-93.

Ephenhuijsen, C.W. 1974. Growing native vegetables in Nigeria. FAO, Rome. pp. 6 and 65-67.

FAO, 1986. Sesame seed production. *FAO Monthly Bulletin of Statistics*, 9: 42.

FAO, 2007. URL: http:/www.fao.org.

FAO. 1999. FAO Production Year Book, 53: 72-73.

Felsenstien J 1985. Confidence limits on phylogenies: an approach using the bootstrap. *Evolution*. 39: 783–791.

Finlay, K.W. and G.N. Wilkinson. 1963. The analysis of adoption in a plant breeding programme. Aust. *J. Agric. Res.*, 14: 742-754.

Freeman, G.H. 1973. Statistical methods for the analysis of genotype environment interaction. *Heredity*, 31: 339-354.

Freeman, G.H. and J.M. Perkins. 1971. Environmental and genotype environmental components of variability VIII. Relation between genotype grown in different environments and measure of these environments. *Heredity*, 27: 15-23.

Gahalain, S.S. 2006. Genetic divergence in rice (*Oryza sativa* L.) genotypes grown in Kumaun Himalaya. *Indian J. Genet.*, 66(1): 37-38.

Gamble, J.S. 1972. A manal of Indian Timbers. Bishen Singh Mahendra Pal Singh, Dehra Dun, pp. 652-654.

Ganesh, S.K. and G. Soundarapandian. 1987. Association studies for stability parameters in short duration varieties of rice (*Oryza sativa* L.). *Heredity*, 25: 36-39.

Ganeshan, S. and P.E. Rajasekharan. 2000a. Conservation of nuclear genetic diversity (NGD) in citrus genepool: Implications in citrus breeding and genetic conservation. In: Higher Tech Citrus Management, Proceedings of International Symposium on Citrus Culture (Nov. 23-27, 1999) held at NRC for Citrus, Nagpur, pp. 85-90.

Gautam, P.L., D.K. Garg, S. Pal, S.K. Malik and U.P. Manglik. 1986. Genotype × environmental interactions and crop improvement. A Review. In: Gupta, P.K. and J.R. Bahal, eds., Genetics and crop improvement, Ragtogi and Company, Subhash Bazar, Meerut, India.

Geetha, S., A.P.M.K. Soundararaj, S. Giridharan and S. Mohandass. 1994. Stability analysis of some medium duration rice genotypes over different nitrogen levels. *Ann. Agri. Res.*, 15(4):391-396.

Gill, K.S. 1990. Prospects for agricultural exports from India. *Agric. Sit. in India*, XLV, 275-280.

Girishkumar, K. 1998. Market assisted selection and mapping quantitative *trait loci* for blast disease resistance in rice (*Oryza sativa* L.). Ph.D. Thesis, Univ. of Agric. Sci., Bangalore.

Gofur, M.A., H.Z. Shafique, D.H. Helati, M. Ibrahim, M.M. Rehman and A. Hakim. 1994. *Bangladesh J. Sci. and Industrial Res.*, 29: 163-171.

GOI, 1990a. Miraculous achievement of Indian agriculture. *Agric. Sit. in Indai*, XLV: 253-255.

GOI, 1990b. All India final estimates of food grains, oilseeds and other crops, 1988-89. *Agric. Sit. in India*, XLV: 876.

Govinda Raju, 1982. Genetic variability, correltion and divergence in *Sesamum indicum* L. M.Sc. (Ag.) Thesis, submitted to Tamil Nadu Agricutlrual University, Coimbatore, India.

Gowda, I.N.D. and A.G. Huddar. 2004. Investigations of processing quality of some mango varieties, hybrids and their blends. *J. Food. Sci. Technol.*, 41(2): 154-159.

Grafius, J.F. 1956. Components of yield in oats: a geometrical interpretation. *Agron. J.*, 48: 419-423.

Griffing, B. and E.W. Lindstorm. 1954. A study of the combining abilities of corn inbreds having varying proportions of corn belt and non-corn belt germplasm. *Agron. J.*, 46: 545-552.

Grubben, G.J.H. 1977. Okra (In) Tropical vegetables and their genetic resources. IBPGR, Rome, pp. 111-114.

Gupta, R.K. and V.S. Brahmachari. 2004. Effect of foliar application of urea, potassium nitrate and NAA on fruit retention, yield and quality of mango cv. Bombai. *The Orissa J. Hort.*, 32(2): 24-27.

Guruprasad, T.R. and Thimaraju, 1989. Seed and spine characters in three jackfruit Artocarpus heterophyllus Lam. types. *Environ. Ecol.*, 7: 511-512.

Habib, A.F. 1971. Cytology of jack. *Mysore J. Agric. Sci.*, 6: 200-202.

Halwart M, Gupta MV 2004. The rice field ecosystem. In: Culture of Fish in Rice Fields. FAO and The World Fish Center, pp. 5–11.

Hanamnaratti, N.G., S.J. Patil, S.K. Nadaf, N. Dronavalli and M. Dronavaili. 1998. Genetic divergence in upland rice (*Oryza sativa* L.) genotypes under low land and upland conditions. *Karnataka J Agric. Sci.*, 11(l): 220-222.

Hartigan, J.A., 1975. Clustering Algorithm. John Wiley & Sons, New York, p. 351.

Hayman, B.I. 1954a. The analysis of variance of diallel tables. *Biometrics*, 10: 235-244.

Hayman, B.I. 1954b. The theory and analysis of diallel crosses. *Genetics*, 39: 789-809.

Hayman, B.I. 1958. The theory and analysis of diallel crosses-II. *Genetics*, 43: 63-85.

Hayward, W.B., 1979. The application of the diallel cross to out breeding species. *Euphytica*, 28: 729-737.

Hazra, P., D. Basu and F.K. Sahu. 2002. Genetic divergence in okra. *Indian J. Hort.*, 59(4): 406-410.

Hedge, S. and B. Vidyachandra. 1998. Yield stability analysis of rice hybrids. *IRRN*, 23(l): 30.

Hedge, S. and C.S. Patil. 2000. Genetic divergence in rainfed rice. *Karnataka J. Agric. Sci.*, 13(3): 549-553.

Hill, J. 1975. Genotype environment interaction a challenge for plant breeding. *J. Agric. Sci.*, 85: 477-493.

Hill, J. and J.M. Perkins. 1969. The environmental induction of heritable changes in *Nicotiana rustica*. Effects of genotype environment interactions. *Genetics*, 61: 661-675.

Hoggag, L.F., M.A. Maksoud and F.M.Z.E.L. Barkouky. 1995. Effect of boron sprays on sex ratio and fruit quality of mango (*Magnifera indica*). *Hindi. Ann. Agri. Sci.*, 40.

Honarnejad, R. 2000. Stability and productivity of Iranian rice cultivars. *Tropenland Wirt*, 10(l): 3-12.

Honarnejad, R., H. Dorosti, M. Salehi and A. Tarang. 1998. Estimation of stability and adaptability of rice cultivars in different environmental conditions. *Seed and Plant*, 13(4): 32-43.

Hussain, M. and Haque, A. 1977. Studies on the physical characteristics of jackfruit. *Bangladesh Hort.*, 5: 9-14.

Hutchinson, J. 1959. The families of flowering plants Vol. I Dicotyledons. Oxford University Press, London, p. 88.

Ibrahim, S.M., A. Ramalingam and M. Subramanian. 1992. Multivariate analysis of genetic divergence over different environments in upland rice. *Oryza*, 29: 141.

Iftekharuddaula, K.M., K. Akter, M.S. Hassan and M.A. Badsha. 2002. Genotype environment interaction in irrigated rice. *Thai. J. Agric. Sci.*, 35(3): 229-236.

Iiyas Ahmed, M. 2003. Hybrid rice; the current global status. In: Winter School on Advances in Hybrid Rice Technology, Eds. B.C. Viraktamath, M. Iiyas Ahmed and B. Mishra, Directorate of Rice Research, Hyderabad.

IRC (International Rice Commission), 2003. Why rice? In: The International Year of Rice 2004-Background, IRCN, p. 52.

IRRI, 1995. Water – A looming crisis. IRRI Manila, Philippines, p. 90.

Irvine, F.R. 1952. Supplementary and emergency food plants of West Africa. *Econ. Bot.*, 6: 23-40.

Ise, K., S. Youquan, Z. Tiande, L. Jishin, S. Kudo, H. Tanno and Y. Sunohara. 2001. Genotype by environment interaction analysis *t1*or rice yield. *Japanese J. Trop. Agric.*, 45: 22-32.

Iyer, C.P.A. and M.R. Dinesh. 1996. Advances in classical breeding and genetics in mango. *Acta Hort.*, 455: 252-267.

Jag Paul Sharma, A.K. Singh, Satesh Kumar and Neerja Sharma. 2008. Genetic divergence studies in okra (*Abelmoschus esculentus* (L.) Moench). *J. Res.*, 7(1): 99-104.

Jaggi, T.K., R.O. Das and D.C. Disen. 1988. Yield study of different genotypes of rice (*Oryza sativa* L.) as influenced by season. *Indian J. Agric. Sci.*, 58(3): 389-392.

Jarrett, F.M. 1959. Studies in Artocarpus and allied genera. *J. Arnold Arbor.* 40: 1-37

Jauhari, O.S. and S.ingh, R.B. 1971. Beal – a valuable fruit. *Indian Hort.*, 6: 9-10.

Jawahar Ram and D.V.S. Panwar. 1970. Intra specific divergence in rice. *Indian J. Genet.*, 30: 1-10.

Jeswani, C.M., B.R. Murthy and R.B. Mehra (1970) Divergence in relation to geographical origin in a world collection of linseed. *Indian J. Genet.*, 30: 11-25.

John Joel, A.J. 1987. Multivariate analysis in sesame (*Sesamum indicum* L.) M.Sc. (Ag.) Thesis Submited to Tamil nadu Agricultural University, Coimbatore, India.

Joshi, A.B. and M.W. Hardas. 1956. Alloploid nature of okra (*Abelmoschus esculentus* (L.) Moench). *Nature*, 178:1198.

Joshi, A.B. and N.L. Dhawan, 1966. Genetic improvement in yield with special reference to self-fertilizing crops. *Indian J. Genet.*, 26: 101-113.

Joshi, A.B., V.R. Gadwal and M.W. Hardas. 1974. Okra In: Evolutionary studies in world crops. Diversity and change in the India Sub-Continent. Cambridge Univ. Press, London, pp. 99-105.

Joshi, A.K. and Y.R. Shukla. 1997. Diversification of vegetable crops in the temperate regions of India. A model. *Punjab Veg. Grower*, 32: 24-27.

Jowett, D. 1972. Yield stability parameters for sorghum in Africa. *Crops. Sci.*, 12: 314-317.

Kalitha, U.C., D.K. Baruah and U.P. Upadhaya. 1993. Bogabordhan, a stable, high yielding, low input traditional variety of Assam, India. *IRRN*, 18(3): 22.

Kalra, S.K., I.S. Yadav and G.C. Shinha. 1994. Screenign of some north Indian mango varieties for processing. *Indian J. Hort.*, 51(2): 136-140.

Kamau, H.N. 1992. Germplasm exploration and collection. *Dinteria*, 23: 50-54.

Kandhola, S.S. and D.V.S. Panwar. 1999. Genetic divergence in rice. *Ann. Biol.*, (Ludhiana), 15(1): 35-39.

Kartha, A.R.S. and Sethi, A.S. (1957). A cold percolation method for rapid gravimetric estimation of oil in small quantities of oil seeds. *Indian J. Agric. Sci.*, 27: 217.

Karthikeyan, T. 2002. Studies on genetic divergence and genotype environment interaction in rice. M.Sc. (Ag.) Thesis, Annamalai Univ., Annamalainagar.

Kiran Patro, T.S.K.K., C. Ravisankar and P.V. Rama Kumar. 2005a. Genetic divergence in bhendi (*Abelmoschus esculentus* (L.) Moench) genotypes. *Andhra Agric. J.*, 52(3 & 4): 507-511.

Knight, R. 1970. The measurement and interpretation of genotype × environment interactions. *Euphytica*, 19: 225-235.

Knight, R.J. 1993. Evaluating important fruit characters in mango germplasm. *Fruits*, 47(1): 25-31.

Krishnankutty, C.N. 1998. Timber price trends in Kerala. *KFRI Research Report*, 160-181.

Kumar, N. 1999. Introduction to Horticulture, 3rd Edition, Rajalakshmi Publication, Nagercoil, Tamil Nadu, pp. 3.1-3.2.

Kumar, R. and Y. Selvaraj. 1990. Fructose 1-6, diphosphatase in ripening mango (*M. indica* L.) fruit. *Indian J. Expt. Bot.*, 28: 284-286.

Kumar, S. and Singh, J.S. 1996. Improvement of jack fruit (*Artocarpus heterophyllus* Lam.) through selection. *Recent. Hort.*, 3: 55-59.

Kurien, T. 1982. A report on project planning. P.G. Diploma, Project Report, Agrl. Engg. Division, College of Horticulture, Kerala Agricultural University, Vellanikara, Trichur.

Lakshminarayana, S., S. Nagy and P.E. Shaw. 1980. Mango. In: Tropical and subtropical fruits: Composition, properties and use. AVI Publishing, West Port, pp. 184-257.

Lawrence, G.H.M. 1973. Taxonomy of vascular plans. Oxford IBH, New Delhi, p. 464.

Ledig, F.T. (1974). Analysis of methods for the selection of tree from wild strands. *For Sci.*, 20: 2-16.

Lefort-Buson, 1987. Heterosis and genetic distance in rapeseed (*Brassica napus* L.). Use of kinship coefficient. *Genome*, 29: 11-18.

Lin, C.S. and M.R. Binns. 1988. A method of analyzing cultivar location year experiment a new stability parameter. *Theor. Appl. Genet.*, 76: 425-430.

Liu K, Muse SV 2005. Power Marker: Integrated analysis environment for genetic marker data; *Bioinformatics*. 21: 2128–2129

Liu, G.Q. 1992. Study of yield stability in new rice varieties in a regional test in North China. *J. Hebel Agrl. Univ.*, 15(l): 8-13.

Lohithaswa, H.C., H.O. Bhushana, D. Basavarajaiah and H.C. Prasanna. 1999. Stability analysis of rice (*Oryza sativa* L.) hybrids. *Karnataka J. Agric. Sci.*, 12(1-4): 48-54.

Luthra, O.P. and R.K. Singh. 1974. A comparison of different stability models in wheat. *Theor. Appl. Genet.*, 45: 143-147.

Mahajan, R.K. and A.S.R. Prasad. 1986. Principal component analysis of genotype environment interaction in rice. *Crop Improv.*, 13(1): 58-62.

Mahalanobis, P.C. (1936). On the generalized distance in statistics. *J. Genet.*, 41:159-193.

Mahalanobis, P.C. 1936. On the generalized distance in statistics. *Proc. Natl. Inst. Sci., India*, 2: 49-55.

Mahapatra, K.C. and S. Das. 1998. An alternative approach to analysis of genotypic adaptation in rice. *Indian J. Genet.*, 58(3): 297-306.

Mahapatra, K.C. and S. Das. 1999. Stability of yield in relation in component traits in rice. *Oryza*, 36(4): 301-305.

Mahapatra, K.C., C.H.P. Mishra and B. Singh. 1996. Assessment of adaptability in early rice varieties. *Indian J. Genet.*, 56(4): 477-485.

Maiti, C.S., Wangchu, L. and Mitra, S.K. 2002. Genetic divergence in jack fruit (*Artocarpus heterophyllus* Lamk.). *Indian J. Genet.*, 62(4): 369-370.

Major Singh, Gyanendra Singh, D.T. Bhutia and R.P. Awasthi. 1995. Genotype environment interaction in rice (*Oryza sativa* L.). *Indian J. Agric. Sci.*, 65(11): 832-833.

Majumder, N.D. and D.N. Borthakur. 1991. Stability for yield in cold tolerant rice. *Oryza*, 33: 11-17.

Majumder, N.D. and D.N. Borthakur. 1996. Phenotypic stability for grain yield and its components in rainfed autumn rice (*Oryza sativa* L.). *Indian J. Genet.*, 56(2): 214-218.

Mamta Kumari and D.N. Choudhury. 2006. Genetic divergence in okra [*Abelmoschus esculentus* (L.) Moench]. *Veg. Sci.*, 33(1): 71-72.

Manamalli, S. 1996. Studies on genetic divergence in greengram (*Vigna radiata* (L.) Wilczek). Ph.D. Thesis, Annamalai University, Annamalainagar, Tamilnadu.

Manay, N.S. and M. Shadaksharaswamy. 1995. Foods: Facts and Principles, Wiley Eastern Ltd., New Delhi, p. 15.

Martin, F.W. 1982. Okra, potential multiple purpose crop for the temperate zones and tropics. *Econ. Bot.*, 36(3): 340-345.

Martin, F.W. and R. Ruberte. 1978. Vegetables for the hot humid tropic Pt. two okra [*Abelmoschus esculentus* (L) Moench]. USDA-ARS- S.R. New Orleans, Louisiana, pp. 1-2.

Mather, K. 1943. Polygenic inheritance and natural selection. *Biol. Revs. Cambridge Phil. Soc.*, 18: 32-64.

Matsuo, T. 1975. Adaptability, stability and productivity of varieties in crop plants. In: T. Matsuo (Ed.) Adaptability in plants. JIBP Synthesis. Univ. Tokyo Press, Tokyo, 6: 173-177.

Mazumdar, B.C. 1975. Physico-chemical analysis of some types of beal (*Aegle marmelos* Correa.) fruit growing in West Bengal. *Indian Agric.*, 19: 295-298.

Mishra, L.K. and A.P. Agarwal. 1992. Stability analysis of yield in rice cultivars. *Adv Plant. Sci.*, 5(1): 197-199.

Mishra, S. and K.C. Mahapatra. 1998. Yield stability of early rice varieties in stratified environments. *Oryza*, 35(2): 109-112.

Mishra, S.B., O.K. Singh and S.P. Sahu. 1994. Studies on genetic divergence in fine land rules of rice in North Bihar. *Ann. Agric. Res.*, 14(3): 276-281.

Mitra, S.K. 1998. Genetic diversity of jackfruit (*Artocarpus heterophyllus* Lamk.). *Indian J. Hort.*, 43(3): 1.

Morishima, H. and H.I. Oka. 1960. The patterns of interspecific variation in the genus Oryza: its quantitative representation by statistical methods. *Evolution*, 14: 153-165.

Mote, U.N., Ajri, D.S. and Tambe, A.B. 1992. An annotated list of pests infesting pomegranate in Central Maharashtra. *Maharashtra J. Hort. Sci.*, 6(1): 96-98.

Muhammed Majeed 1994. Citrin-A revolutionary herbal approach to weight management. New Editions Publishing Company, Burlingane, p. 8-14.

Murthy (1965) Heterosis and combining ability in relation to genetic divergence in flue cured tobacco. *Indian J. Genet.*, 25: 46-56

Murthy, B.R., G.S. Murthy and M.V. Pavats. 1962. Studies in quantitative inheritance in *Nicotiana tobacum*. Components of genetic variation for flowering time, leaf number, grade performance and leaf burn. *Zuchter*, 361-369.

Muthulakshmi, P., Sarah T. George and K. Lila Mathew (1999). Morphological and biochemical varition in different sex forms of Kodampuli (*Garcinia gummi-gutta* L.). *J. Trop. Agric.*, 37: 28-31.

Nagaraj, G. 1990. Biochemical quality of oilseeds. *J. Oilseeds Res.*, 7: 47-55.

Naidu, B.S., M.S. Badraiah, T.A. Khan and Jayakumar. 1980. IET 2490 and IET 2730 have high grain yield stability. Curr. Res. Uni. of Agric. Sci., *Bangalore*, 9: 148-149.

Naik, K.C., Bhujanga Rao, C. and Raman V.S. 1959. Problems in crop improvement in mango – proceedings of International Sympoisum on the origin, Cytogenetics and Breeding of tropical fruits. *Indian J. Hort.*, 16(1): 159-167.

Nandini, P.V. and O.B. Elizabeth. 2002. Quality parameters of selected mango cultivars. *J. Food Sci. Technol.*, 39(4): 379-383.

Nanita Devi, H., N.B. Singh*, M.R.K. Singh and P.R. Sharma. 2006. Stability analysis of selected rice genotypes of varying spacing and sowing dates under rainfed lowland condition in Manipur Valley, *Oryza*, 43(1): 20-24.

Narasimman, M., 1929. A note on the pollination of black and greengram in the Godawari district. *Agri. J. India*, 24: 397-401.

Narasinhan, P. 1990. Bread frit and jackfruit. In: Nagy, S., Shaw, P.E. and Wordowski, W.F. (ed.) Fruits of tropical and subtropical origin: Composition properties and uses. Florida Science Source Ltd., Florida.

Narendrakularni and A. Gangaram. 1988. Genetic changes and phenotypic stability in mutants of rice variety samba mashuri. *Oryza*, 35(4): 322-324.

Narine, A.K. 1894. The flowering plants of Western India. W.H. Allen and Co., London, p. 309.

Natarajan, C., K. Thiyagarajan and R. Rathnaswamy, 1988. Association and genetic diversity studies in greengram [*Vigna radiata* (L.) Wilczek]. *Madras Agric. J.*, 75(7-8): 238-245.

Nelson, C.H., H.J. Nieschin, M.E. Daxenbichler, I.A. Wolff and R.E. Jr. Perdue. 1961. A search for new fiber crops. III laboratory-scale pulping studies. *Tappi.*, 44: 319-325.

Nuzhatfaroogui, S.A., Ranade, P.U. Sane, and N. Farroqui, 1998. RAPD profile variation amongst province of neem. *Biochem. Mol. Bio. Int.*, 45(5): 931-939.

Palaniswamy, K.P., C.R. Muthukrishnan and K.G. Shanmugavelu. 1974. *Indian F. Packer*, 28: 12-19.

Pande, K. and D.P. Ghora. 1987. Genetic divergence in rice. *Oryza*, 24: 353-357.

Pande, K., S. Singh and O.N. Singh. 2006. Stability of rice (*Oryza sativa* L.) varieties for boro season of eastern India. *Indian J. Genet.*, 66(3): 191-195.

Panhwar, F. 2005. Genetics breeding and selection of mangoes in Pakistan: In http://www.chemlin.com.

Perkins, J.M. 1970. Environmental and genotype environmental components of variability. VI Diallel sets of crosses. *Heredity*, 25: 29-40.

Perkins, J.M. and J.L. Jinks. 1968. Environment and genotype environmental components of variability III multiple line and crosses. *Heredity*, 23: 339-356.

Phadnis, N.A. 1974. Pomegranate for dessert and juice. *Indian Hort.*, 19(3): 9-14.

Philomina, D (2000). Genetic analysis of one-parent families for variability, diversity, stability and propagation techniques in Neem (*Azadirchta indica* A. Juss). Ph.D.Thesis, TNAU, Coimbatore.

Plaisted, R.L. and L.C. Peterson. 1959. A technique for evaluating the ability of selection to yield consistently in different locations and seasons. *Amer. Potato J.*, 36: 381-385.

Pradhan, K. and A. Roy. 1990. Genetic divergence in rice. *Oryza*, 27: 415-418.

Prasad, R.S. 1998. Genotype environment interaction for yield related traits and identification of QTL linked to salt tolerance in rice (*Oryza sativa* L.) DH population. M.Sc. (Ag.) Thesis, Univ. of Agric Sci., Bangalore.

Purseglove, J.W. 1974. Tropical crops: Dicotyledons. Vol.1 and 2 combined, Longman, London.

Purseglove, J.W., 1968. Tropical crops-dicotyledons, 3rd Reprint, 1974. Longman, London.

Rai, B. 1979. Heterosis breeding. Agro-biological Publications, Delhi, pp. 184.

Rajasekaran, A. 2006. Studies on genetic divergence of Indica and Japonica rice (*Oryza sativa* L.) genotypes. M.Sc. (Ag.) Thesis, Annamalai University, Annamalainagar.

Ram, J., O.P. Jain and B.R. Murthy. 1970. Stability performance of some varieties and hybrid derivatives of rice under high yielding varieties programme. Ann. Cog. Indian Soc. Agric. Stat., Patna.

Raman, V.S., Kesavan, P.C., Manimekalai, G., Alikhan, W.M. and Rangaswami, S.R. 1963. Cytological studies in some tropical fruit plants. *South Indian Hort.*, 11(3-4): 27-33.

Ramanujam, S., A.S. Tiwari and R.B. Mehra, 1974. Genetic divergence and hybrid performance in mungbean. *Theor. Appl. Genet.*, 45: 211-214.

Ranga Rao, V., M. Ramachandran and Jawahar Ram Sharma. 1980. Multivariate analysis in genetic divergence in safflower. *Indian J. Genet.*, 40: 72-89.

Rao, C.R. 1952. Advanced statistical methods in biometrical research. Wiley and Sons, New York.

Rao, C.R. 1958. Bengal anthropometric survey. 1945. A statistical study. *Sankhya*, 19: 201-408.

Rao, S.S., A. Bhatnagar, R. Verma, T.D. Pandey and K.L Nandeha. 1996. Stability analysis of 13 early –duration upland rice genotypes in Bastar plateau zone, Madhya Pradesh, India. *IRRN*, 21(1): 24-25.

Rao, T.P. and P. Gomathinayagam. 1997. Genetic divergence in semi-dry rice under different environments. *Madras Agric. J.*, 84(6): 314-317.

Rather, A.G., M.A. Zarger and F.A. Sheikh. 2001. Genetic divergence in rice (*Oryza sativa* L.) under temperate conditions. *Indian J. Agric. Sci.*, 71(5): 344-345.

Rathinaswamy, R. and D. Jagatheesan. 1984. Selection of superior early generation crosses in Sesamum indicum L. based on combing ability study. *Zeitschrift for Pflanzenziichutung*. 93: 184-190.

Ratho, S.N. 1984. Genetic divergence in scented varieties of rice. *Indian J. Agric. Sci.*, 54: 437 -440.

Reddy, J.N. and D. Chaudhary. 1991. Stability for grain yield and its components in rice. *Oryza*, 28(2): 295-299.

Reddy, J.N., D. Pani and J.K. Roy. 1998. Genotype x environment interaction for grain yield in lowland rice cultivars. *Indian J. Genet.*, 58(2): 209-213.

Reddy, J.N., R.N. De, A.V.S. Rao. 2002. Genetic divergence in low land rice under intermediate water (0 50 cm) depth. *Indian Agric.*, 46: 37-44.

Reddy, O.U.K. 1986. Studies on variability, character association and genetic divergence in *Sesamum indicum* L. M.Sc. (Ag.) Thesis submitted to Tamil Nadu Agricultural University, Coimbatore, India.

Reddy, Y.S. and P.V.R. Kumar. 1996. Stability analysis for yield in rice (*Oryza sativa* L.). *New Bot.*, 23: 1-4.

Rendle, A.B. 1971. The classifiation of flowering plants Vol. II – Dicotyledons, Cambridge University Press, London, pp. 51-55.

Roy, A. and D.V.S. Panwar. 1993. Genetic divergence in some rice (*Oryza sativa* L.). *Ann. Agric. Res.*, 14(3): 276-281.

Roy, A. and D.V.S. Panwar. 1994. The stability of some Indian and exotic genotypes of rice. *Ann. Agric. Res.*, 15(2): 147-151.

Sagar, V.R. and D.S. Khurdiya, 1996. Effect of ripening on physico-chemical characteristics on Dasheri mango. *Indian Fd. Packer*, 50(5): 16-23.

Saini, S., G.S. Sidhu, Kumar and T.S. Coomer. 1976. Phenotypic stability for yield of some rice varieties. *Indian J. Agric. Sci.*, 44(10): 644-647.

Sammadar, N. 1990. Jack fruit. Fruits: Tropical and Subtropical (eds. Bose, T.K. and Mitra, S.K.) Naya Prakash, Calcutta, pp. 84-90.

Samonte, S. and J.E. Hernandez. 1991. Genotype × environment interactions of irrigated: stability adaptability analysis. *Philippines J. Crop Sci.*, 16(1): 7-4.

Santapau, H. 1966. Common trees N.B.T. Publications, New Delhi, pp. 11-13.

Sanyal, D., R. Roychoudhury, R.S. Dhua and S.K. Mitra. 1991. *Indian Food. Packer*, 45: 29-33.

Sarah, T.G. (1998) The fruit that fights fat. Indian Express, Science Express Supplement 09 June 1998, p. 4.

Sarathe, M.L., K.K. Sharma and M.N. Srivastava. 1996. Study of yield attributes and heritability in some varieties of rice (*Oryza sativa* L.). *Indian J Agric. Sci.*, 39: 925-929.

Sarawgi, A.K. and M.N. Shrivastava. 1996. Genetic divergence in rice under irrigated and rainfed conditions. *Adv. Plant Sci.*, 9(2): 93-100.

Sardana, S. and B. Sasikumar. 1990. Phenotypic stability for grain yield in promising lines of rice (*Oryza sativa* L.). *Indian J. Agric. Sci.*, 60(8): 541-542.

Sarma, M.K., R.K. Agarwal and A.K. Richarida. 1997. Genetic diversity in rice under rainfed upland and irrigated eco system. *Oryza*, 34: 19-24.

Sarma, R., Mohan, N.K. and Nath, J.C. 1997. Evaluation of yield potential and physico-chemical composition of some local types of jackfruit to Assam. *Hort. J.*, 10: 23-30.

Satyabalan, K. (1993a). The Coconut Palm. Botany and Breeding. Published by Asian and Pacific Coconut Community. p. 3.

Satyabalan, K. (1993b). The Coconut Palm. Botany and Breeding. Published by Asian and Pacific Coconut Community. p. 77.

Satyabalan, K. (1993c). The Coconut Palm. Botany and Breeding. Published by Asian and Pacific Coconut Community. p. 123.

Savello, P., F.W. Martin and J.M. Hill. 1980. Nutritional composition of okra seed meal. *Agric. Food Chem.*, 28: 1163-1166.

Sawant, D.S., S.L. Patil and B.B. Jadhav. 1997. Stability analysis for seed yield in early rice genotypes. *Ann. Agric. Res.*, 18(2): 211-213.

Sawant, D.S., S.L. Patil, B.B. Jadhav and S.G. Bhave. 1996. Genetic divergence, character association and path analysis in rice. *J. Maharashtra Agrl. Univ.*, 20(3): 412-414.

Sen, N.K. and M.K. Jana, 1963. Genetics of blackgram (*Phaseolus mungo* L.). *Genetica*, 34: 46-57.

Shadakshari, Y.G., H.M. Chandrappa, R.S. Kulkarni and H.E. Shashidhar. 2001. Genotype environment interaction in lowland rice, genotype of hill zone of Karnataka. *Indian J. Agirc. Sci.*, 61(4): 350-352.

Shanthakumar, G., R.S. Kulkarni and R.C. Jagadeesha. 1997. Stability analysis in rice (*Oryza sativa* L.) over different seasons for yield and its components. *Karnataka J. Agric. Sci.*, 10(1): 67-70.

Shanthi, P. and J. Singh. 2001. Genetic divergence for yield and its components in induced mutants of Mashuri rice. *Res. on Crop* 2(3): 390-392.

Sharma, 1964. *Jackfruit. Proc. Ind. Acad. Sci.*, 60: 380-393.

Sharma, M.R. 1962. Morphological and anatomical investigations on Artocarpus. *Proc. Indi. Acad. Sci.*, 56: 243-258.

Sharma, S.K. Singh, A.K. and Singh, O.P. 2005. A study on genetic variability and germplasm evauation in jackfruit (*Artocarpus heterophyllus* Lamk.). *Adv. Plant Sci.*, 18(11): 549-55.

Sharma, T.R. and S.N. Mishra, 1997. Genetic divergence and character association studies in cowpea. *Crop Res.*, 13: 109-114.

Shi, C.H. and Y.G. Yu. 2000. Genotype × environment interaction effect and genotypic correlation for nutrient quality traits of Indian rice (*Oryza sativa* L.). *Indian J. Agric. Sci.*, 70(2): 85-89.

Shibamoto, T. and C.S. Tang. 1990. Minor tropical fruit mango, papaya, passion fruit and guava. In: Morton, I.D., Mecheod, A.J. (eds.) Food Flavours: Part C. The Flavour of Fruit. Elsevier, Amsterdam, pp. 221-234.

Shiv, D., S.C. Mani and S. Datt. 2003. Genetic divergence in elite genotypes of basmati rice. *Indian J. Genet.*, 63(1): 73-74.

Shivana, S. 1989. Stability analysis of yield and components in medium late maturing genotype of rice (*Oryza sativa* L.). M.Sc. (Ag.) Thesis, Univ. Agric. Sci., Bangalore.

Shivaprakash, N. 1987. Assessment of genotype environment interactions of extra early selection in rice, (*Oryza sativa* L.). M.Sc. (Ag.) Thesis, Univ. Agric. Sci., Bangalore.

Shukla, G.K. 1972. Some statistical aspects of partitioning genotype × environment components of variability. *Heredity*, 29: 237 245.

Shukla, S.K. and M.P. Pandey. 2003. Genetic diversity in elite TGMS lines. *Indian J Genet.*, 63(1): 71-72.

Silva FG da, A. Fijho CJ da and J.N. Tabosa. 1995. Grain yield stability of irrigated rice in Alagoas and pernambuco state. *Pesquisa Agropacuaria Brasileria*, 30(3): 347-351.

Singh, A.K., B.P. Singh and C.B.S. Rajput. 1990. Studies on correlation between physico-chemical properties of fruit in mango (*Magnifera indica* L.). *Res. and Dev. Reporter*, 7: 12-14.

Singh, B.K. and A.K. Singh. 2008. Genotype clustering in okra (*Abelmoschus esculentus* (L.) Moench) using D^2 statistics. *Hort. J.*, 21(1): 34-36.

Singh, B.N. 2002. Rice in India and global rice scenario. In: Training manual of summer school on genetic evaluation and utilization (GEU) in rice improvement held between 6th May to 4th June 2002. Central Rice Research Institute, Cuttack, pp. 1-12.

Singh, I.S. and Srivastava, A.K. 2000. Genetic variability in jack fruit IPGRI – News letter for *Asia the Pacific and Oceania*, 31: 22-23.

Singh, N. and B.M. Gangwar. 1985. Testing the commercial significance of some mid-season mango (*M. indica* L.) cultivars for genetic plains of Northern India. *Punjab Hort. J.*, 25: 12-18.

Singh, N.K., V.K. Sharma and P.B. Iha. 1995. Genotype×environment interaction for yield and related traits in Indica rice. *J. Agric. Res.*, 72: 121-125.

Singh, R.B. 1983. Studies on genetic variability in rice. *Madras Agric. J.*, 70: 436-440.

Singh, R.B. and M.P. Gupta. 1968. Multivariate analysis of divergence in upland cotton. *Indian J. Genet.*, 28: 151-157.

Singh, R.K. and B.D. Choudhari. 1985. Biometrical methods in quantitative genetic analysis. Revised ed. Kalyani Publishers, New Delhi.

Singh, R.S. and S.S. Bains. 1968. Genetic divergence for ginning out turn and its components in upland cotton (*G. hirsutum* L.) varieties. *Indian J. Genet.*, 28: 262-268.

Singh, R.S., P.P. Singh, D.M. Mourya and O.P. Verma. 1987. Genetic divergence among low land rice cultivars. *Indian J. Genet.*, 47(1): 11-14.

Singh, S. Krishnamoorthy, S. and Kaytal, S.L. 1967. Fruit Culture in India, ICAR, New Delhi, p. 27.

Singh, U., U.P. Singh and I.B. Singh, 1976. Discriminant function techniques for the improvement of grain yield in blackgram. *Trop. Grain. Legume Bull.*, 6: 15-16.

Singh, U.K., S.B. Mishra and R. Thakur. 1999. Genetic divergence in boro Rice. *Oryza*, 36(1): 76-77.

Sinha, S.K. and S. Biswas. 1987. Adaptability of rice varieties during boro season of West Bengal. *Indian Agric.*, 31(1): 15-18.

Smit, M. and K.C. Mahapatra. 1998. Phenotypic stability for grain yield in rice. *Oryza*, 35(3): 269-271.

Sokal, R.R. 1965. Statistical analysis in systematic. *Biol. Rev.*, 40: 387-391.

Soni, D.K., A.K. Sarawgi and N.K. Rastogi. 1999. Genetic divergence in traditional rice accessions of Madhya Pradesh, India. *Oryza*, 36(2): 118-120.

Sood, B.C. and G.C. Saini, 1989. Stability for yield and maturity in mash [*Vigna mungo* (L.) Hepper]. *Indian J. Gent.*, 49: 309-311.

Sree Krishna, B. 2005. Studies on genetic diversity and stability analysis in rice (*Oryza sativa* L.). M.Sc. (Ag.) Thesis, Annamalai University, Annamalainagar.

Srinivasan, K. 1980. Muttam varikka – A promising jack fruit variety. *Agric. Res. J. Kerala*, 8: 51-52.

Subramanyan, M.D. and Iyer, C.P.A. 1981. Studies on variability, heritability and genetic advance in Carcia papaya cultivars. *S. Indian Hort.*, 29: 167-174.

Sudheer, K.S., S. Vinod Kumar and V. Narsimha Reddy. 1996. Phenotypic stability and rainfed rice (*Oryza sativa* L.). *Indian J. Agric. Sci.*, 66(12):705-707.

Suman, K., M.R., R.S. Kulkarni and S. Ramesh. 1999. Identification of stable rice genotypes for late planting situations under tankfed areas. *Mysore J. Agric. Sci.*, 33: 347 354.

Suresh Babu, K.V. 1987. Cytogenetical studies in okra [*Abelmoschus esculentus* (L) Moench]. Ph.D. Thesis, Univ. Agrl. Sci., Bangalore.

Suresh Pal, 1989. Stagnant production and changing production instability of oilseeds in India: 1950-1984. *Agric. Sit. in India, XLIV*: 1-26.

Swaminathan, M.S. 1989. Role of rice in global food security. *Oryza*, 26: 1-26.

Tai, C.P. 1971. Genotypic stability analysis and its application to potato regional trials. *Crop. Sci.*, 11: 184-190.

Tessy, J. 1982. Floral biology and fruit development in varikka and koozha types of Jack. M.Sc. (Ag.) Thesis, Kerala Agricultural University, Thrissur, India, p. 101.

Thangavelu, M.S. and S. Rajasekaran. 1983. Correlation and path coefficient analysis in sesamum. *Madras Agric. J.*, 70: 109-113.

Thirugnanakumar, S., 1991. Seed genetics in relation to yield in sesame (*Sesamum indicum* L.). Ph.D. Thesis, Tamil Nadu Agricultural University, Coimbatore.

Thirugnanakumar, S., Inasi, K.A. and Sree Kumar, K. (2006a). Genetic variability and correlation studies among fruit characters of gamboge tree (*Garcinia gummi-gutta* L.). *Indian J. Forestry*, 29(2): 139-143.

Thirugnanakumar, S., Inasi, K.A. and Sree Kumar, K. (2006b). Studies on genetic diversity among the naturally distributed populations of gamboge tree (*Garcinia gummi-gutta* L.). *Indian J. Forestry*, 29(4): 429-434.

Timothy, D.H. 1963. Genetic diversity in heterosis and use of exotic stocks in maize at Columbia. Proc. Statistical Genet. Pl. Breed. Symp. (Raleigh): *NAS-NRC, Publ.* 982: 581-893.

Tomoda, M., K. Shimada, Y. Saito and M. Sugi. 1980. Plant mucilages : 26, Isolation and structural features of a mucilage, okra-mucilage F, from the immature fruits of [*Abelmoschus esculentus* (L) Moench]. *Chem. Pharm. Bull.*, 28: 2933-2940.

Trehan, K.B., A.V. Rao, S.K. Metha, H. Chand, H.N. Sharma and S.K. Baijal. 1974. Genetic divergence in sesame. *Indian J. Agric. Sci.*, 44: 208-212.

Upadhya, M.K. and B.R. Murthy. 1970. Genetic divergence in relation to geographical distribution in pearl millets. *Indian J. Genet.*, 30: 704-715.

Vasugi, 2010. Studies on morphological and molecular characterization and conservation of mango germplasm. Ph.D. Thesis, Annamalai University, Chidambaram.

Veeraraghavadattam, D., Jeeva, S. and Rabindran, R. 1996. Scientific Fruit Culture. Suri Associates, Coimbatore, p. 129.

Venkatasubbhiah, N. 2006. Stability analysis in rice (*Oryza sativa* L.) M.Sc. (Ag.) Thesis, Annamalai University, Annamalainagar.

Venkatesan, C.K. 2004. Studies on genetic divergence and stability analysis in rice (*Oryza sativa* L.). M.Sc. (Ag.) Thesis, Annamalai Univ., Annamalainagar.

Verdicourt, B., 1970. Studies in the Leguminosae, Papilionidae for the Flora of Tropical East Africa. *Kew Bull.*, 24: 507-569.

Vergara, S.B. 1985. Oryza sativa. In: CRC Handbook of Flowering, Vol. III, ed. A.H. Halevy, CRC Press, Inc. Boca Raton, Florida, pp. 435-441.

Verma, M.M. and D.S. Virk. 1978. Limitations of conventional regression analysis for varietal adoption. *Crop. Improv.*, 10(2): 67-78.

Verma, M.M. and G.S. Chahal. 1978. Limitations of conventional regression -analysis, proposed modification. *Theor. AppL Genet.*, 3: 89-91.

Verma, S. and G. Kanta, 1997. Divergence analysis in blackgram over cropping system. *Crop Improv.*, 24: 97-100.

Vijendra, L.D., N. Shumugavalli and P. Veluswamy. 1995. Phenotypic stability in rice. *Madras J. Agric.*, 82(11): 569-571.

Viruel, M.A., P. Escribano, M. Barbieri, M. Ferri and J.I. Hormaza. 2005. Finger printing, embryo type and geographic differentiationin mango (*Magnifera indica* L., Anacardiaceae) with micro satellites. *Mol. Breed.*, 15: 383-393.

Vivekanandan, P. and Sukanyasubramaniyan. 1993. Genetic divergence in rainfed rice. *Oryza*, 30: 60-62.

Wallace, B. 1963. Models of reproduction and their genetic consequences in statistical genetic and plant breeding. Symp. Raleigh, pp. 3-20.

Wilcox, J.B. and C.P. Wilsie, 1964. Estimated general combining ability, specific combining ability effects in crosses among nine clones of alfalfa. *Crop. Sci.*, 4: 375-377.

Williams, W., 1964. Genetical principles and plant breeding. Blackwell Scientific Publications, Oxford, p. 504.

Woodroof, J.G. 1927. Okra. *Georgia Exp. Stn. Bull.*, 145: 164-165.

Woolfe, M.L., M.F. Chaplin and G. Otchere. 1977. Studies on the mucilage extracted from okra fruits (*Hibiscus esculentus*) and baobab leaves (*Adansonia digitata*). *J. Sci. Food Agric.*, 28: 519-529.

Wricke, G. 1962. Uber eine methode zur Erfassung der okologischen strenbreite in feldrersuchen. *Z Pflzucht.*, 47: 92 96.

Yang Chang, I.H.N., J.U.N. Yang Sae, J.Y. Pyeong, C.H. Chune, S.Y. Boum, C.I. Yang, S.J. Yang, Y P. Jeoung, H.C. Choi and Y.B. Shin. 2001. Genotype × environment interaction of rice yield in multilocation trails. *Korean J. Crop Sci.*,46(6): 453-458.

Zaied, N.S., S.A.A. Khafagy and M.A. Salah. 2007. Evaluation of some mango species by fruit characters and fingerprint. *Res. J. Agri. and Biol. Sci.*, 3(4): 316-320.

Ziauddin Ahmad, R.P., Kattiyar and Radhey Sharma 1980. Genetic divergence in triticale. *Indian J. Genet.*, 40: 35-39.

Zobel, B. and Talbert, J.I. 1986. Applied Forest Tree Improvement. New York, John Wiley and Sons, p. 505.

Other Publications on Plant Breeding and Genetics

S.No.	Title	Author	ISBN	Year
1	Abiotic Stress and Physiological Process in Plants	A. Bhattacharya	9789385516696	2017
2	Abiotic Stress Tolerance in Crop Plants: Breeding and Biotechnology	Roy, Bidhan	9788189422943	2009
3	Agriculture Bioinformatics	Keshavachandran, R.	9789383305421	2015
4	Agro-Informatics	Vanitha, G & M.Kalpana	9789380235714	2011
5	An Introduction to Intellectual Property Rights	Pathak, Manju	9789383305124	2014
6	Applied Computational Biology and Statistics in Biotechnology and Bioinformatics (Set of 2 Vols.)	Roy, A.K.	9789380235929	2012
7	Basics of Mutation Breeding	Thirugnanakumar, S.	9789383305193	2014
8	Biochemistry,Molecular Biology and Biotechnology: Instant Notes	Gajera, H.P.,	9789383305520	2015
9	Bioinformatics in Agriculture: Tools and Applications	Balakrishnan, M. & Dam Roy ed.	9789381450925	2014
10	Biological Sciences: Innovations and Dynamics	Sinha, Rajeshwar P.	9789383305872	2015
11	Biometrical Methods in Horticultural Science	Nirmal Sharma, V.K. Wali, Parshant Bakshi	9789385516504	2016
12	Biotechnology in Horticulture: Methods and Applications	Peter, K.V.	9789381450918	2013
13	Biotechnology in India: Initiatives and Accomplishments	Niladri, Bag	9789385516252	2016
14	Biotechnology: Practical Manual Series Vol 04	Thara, K.M.	9788190851237	2009
15	Breeding and Biotechnology of Flowers: Set of 2 Vols. (Set Price)	A.K.Singh	9789383305612	2014
16	Breeding and Biotechnology of Flowers: Vol 01: Commercial Flowers	Singh, A.K.	9789383305353	2014
17	Breeding and Biotechnology of Flowers: Vol 02: Garden Flowers	Singh, A.K.	9789383305407	2014
18	Breeding and Protection of Vegetables	Rana, M.K.	9789380235493	2011
19	Breeding of Horticultural Crops: Principles and Practices: 2nd Revised & Expanded ed.	Kumar, N.	9789383305773	2015
20	Breeding, Biotechnology and Seed Production of Field Crops	Roy, Bidhan	9789381450680	2013
21	Developments in Physiology,Biochemistry and Molecular Biology of Plants Vol 01	Bose, Bandana & A.Hemantaranjan: ed.	9788189422028	2005
22	Developments in Physiology,Biochemistry and Molecular Biology of Plants Vol 02	Bose, Bandana & A.Hemantaranjan: ed.	9788189422929	2008
23	DNA: A Bridge Between Biochemistry and Biotechnology	Dhruve, J. J.	9788189422240	2008
24	Emerging Technologies of the 21st Century	Roy, A.K.	9789383305339	2015
25	Experimental Biotechnology: Practical Manual Series 06	Dutta, Sunita & Abhijit Dutta	9789380235721	2011
26	Flower Crops: Cultivation and Management	Singh, A.K.	9788189422356	2006
27	Flowering Trees: Vol.12. Horticulture Science Series	Valsalakumari, P.K.	9788189422509	2008
28	Flowers for Trade: Vol.10. Horticulture Science Series	Sheela, V.L.	9788189422516	2008
29	Fruit Breeding	Dinesh, M.R.	9789383305513	2015

30	Genomics and Genetic Engineering	Satya, Pratik	9788189422776	2007
31	Handbook of Genetics and Biotechnology: Revised and Expanded 2nd ed.	Tomar, R.S.	9789383305445	2014
32	Heterosis Breeding in Vegetable Crops	Rai, Nagendra	9788189422035	2006
33	Innovative Horticulture	Arunkumar, K.	9788189422738	2008
34	Intellectual Property Rights Demystified	Ramkumar, Mu.	9788189422875	2008
35	IPR: Drafting,Interpretation of Patent Specifications and Claims	Rathore, N.S.: Ed.	9789381450819	2013
36	Laboratory Manual of Biochemistry: Methods and Techniques	Sengar, R.S.	9789383305025	2014
37	Laboratory Manual of Microbiology: Practical Manual Series: 05	Roy, A.K.	9789380235189	2010
38	Merging Plant Breeding with Crop Biotechnology	Yasin, J.K.:Ed.	9789381450598	2012
39	Microbes for Plant Stress Management	D. Joseph Bagyaraj & Jamaluddin	9789385516658	2017
40	Microbes for Restoration of Degraded Ecosystems	D. Joseph Bagyaraj & Jamaluddin	9789385516665	2017
41	Modern Biotechnology and Its Applications (Set of 2 Vols.)	Behera, K.K.	9789381450833	2013
42	Molecular Biology and Biochemistry	Puttaraju, H.P.	9788189422257	2008
43	Molecular Markers and Plant Biotechnology	Tomar, R.S.	9789380235257	2010
44	Nanotechnology in Agriculture	Subramanian, K.S.	9789383305209	2015
45	Nitrogen Use Efficiency in Plants	Jain, Vanitha & P. Ananda Kumar	9789380235738	2011
46	Objective Genetics	Thirugnana Kumar S. & A. Anandan	9789383305605	2015
47	Physio-Biochemistry and Biotechnology of Vegetable Crops	Rana, M.K.	9789380235318	2011
48	Physiological Disorders of Fruit Crops	Savreet, Sandhu & Bikramjit Singh Gill	9789381450581	2013
49	Phytochemical Techniques	Raaman, N.	9788189422301	2006
50	Pigeonpea Hybrids and Their Production	Tikle, A.N.	9789383305957	2015
51	Plant Secondary Metabolities	Shukla, Y.M. & R.Bhatnagar	9788190851220	2009
52	Plant Taxonomy and Biosystematics: Classical and Modern Methods	Rana, T.S.	9789383305414	2014
53	Plant-Microbe Interactions	Ramasamy, K. and K.Kumar	9789383305834	2015
54	Practical Plant Biotechnology and Genetics	Rani, Archana	9789383305995	2015
55	Propagation of Horticultural Crops: Vol 06 Horticulture Science Series	Rajan, S. & B.L.Markose	9788189422486	2007
56	Propagation of Horticultural Plants: Arid and Semi-Arid Regions	Singh, R.S. & R.Bhargava	9789383305254	2014
57	Quantitative Genetics and Crop Breeding	Thirugnanakumar, S.	9789380235981	2012
58	Radio Tracer Techniques for Agriculturists and Biologists	Golakiya, B.A.	9788189422974	2008
59	Recent Trends in Horticultural Biotechnology: In 2Vols	Keshavachandran, R.	9788189422592	2007
60	Rejuventaion in Fruit Trees: Tropical and Subtropical	Kumar, Rajesh	9789385516436	2016
61	Techniques in Molecular Diagnostics	Kandarpa Kr. Saikai	9789385516528	2016
62	Turning Plants Into Medicines: Novel Approaches	T, Parimelazhagan	9789381450468	2013
63	Vegetable Crops: Genetics Resources and Improvements	Singh,D.K. & H.Choudhary	9789380235509	2012
64	Vegetable Crops: Vol 04. Horticulture Science Series	Gopalakrishnan, T.R.	9788189422417	2007